BEAUTIFUL GAME THEORY

# BEAUTIFUL GAME THEORY

## How Soccer Can Help Economics

IGNACIO PALACIOS-HUERTA

PRINCETON UNIVERSITY PRESS
Princeton and Oxford

Published by Princeton University Press, 41 William Street,
Princeton, New Jersey 08540
In the United Kingdom: Princeton University Press, 6 Oxford Street,
Woodstock, Oxfordshire OX20 1TR
press.princeton.edu

Fourth printing, first paperback printing, 2016
Paperback ISBN: 978-0-691-16925-5
The Library of Congress has cataloged the cloth
edition of this book as follows:
Palacios-Huerta, Ignacio
Beautiful game theory : how soccer can help economics / Ignacio
Palacios-Huerta.
pages cm
Summary: "A wealth of research in recent decades has seen the economic approach to human behavior extended over many areas previously considered to belong to sociology, political science, law, and other fields. Research has also shown that economics can illuminate many aspects of sports, including soccer. *Beautiful Game Theory* is the first book that uses soccer to test economic theories and document novel human behavior. In this brilliant and entertaining book, Ignacio Palacios-Huerta illuminates economics through the world's most popular sport. He offers unique and often startling insights into game theory and microeconomics, covering topics such as mixed strategies, discrimination, incentives, and human preferences. He also looks at finance, experimental economics, behavioral economics, and neuroeconomics. Soccer provides rich data sets and environments that shed light on universal economic principles in interesting and useful ways. Essential reading for students, researchers, and sports enthusiasts, *Beautiful Game Theory* is the first book to show what soccer can do for economics" —Provided by publisher.
Includes bibliographical references and index.
ISBN 978-0-691-14402-3 (hardback)
1. Economics—Psychological aspects. 2. Soccer—Economic aspects.
3. Game theory. I. Title.
HB74.P8.P27 2014
330.01'51932—dc23 2014003345
British Library Cataloging-in-Publication Data is available
This book has been composed in Baskerville 120 Pro & Raleway
Printed on acid-free paper. ∞
Printed in the United States of America
10 9 8 7 6 5 4

*To Ana, Ander, and Julene*
*To María Luz, Javi, Patxo, Anton, Jon,*
*and to the memory of my father José Antonio*

# CONTENTS

BEAUTIFUL GAME THEORY

# INTRODUCTION

Economy is the art of making the most of life.
—George Bernard Shaw

Politics and high-profile sport, like religion,
are about the whole of life.
—Reverend Michael Sadgrove (letter to Paolo Di Canio)

Toward the end of the 2001–2 soccer season, a group of young members of the Brigade of Martyrs of Al Aqsa met with a reporter from the *Sunday Times* of London in the depths of a basement in the Palestinian territory of Gaza. The candidates to martyrdom were talking to the journalist about the pain of their people, their wish to die rather than live like slaves, and their dreams of hearing Israeli mothers mourn their sons' deaths, when another member of the group stormed into the basement. "Manchester United, 5," he announced, "West Ham United, 3. Beckham scored two goals!" The terrorists exploded in joy, shouting "Allah Akbar!" [God is great!].[1]

We hardly need to be reminded of the unique place that soccer occupies in the world today. Things, however, were very different one century ago. As González (2008) notes, the accelerated industrialization of the 19th century bequeathed to the 20th century two mass phenomena that were oddly intertwined: Marxism and soccer. Both were born out of migration from rural areas to the cities and the alienation of the new proletariat. Marxism proposed the socialization of the means of production and the hegemony of the working class as a solution. Soccer proposed a ball, 11 players, and a flag. One century later, there is no doubt about which proposal was more attractive.

Of the three ingredients that soccer offers, the most essential to its success is neither the ball nor the players but the flag. I shall clarify this point. Before the masses were dispossessed, sporting ideals were based on the hero. The great athlete, model of virtue, embodied collective aspirations. In continental Europe, this model was the case until well into the 20th century. It is remarkable that the two oldest sports dailies in Europe, *La Gazzetta dello Sport* (1896) and *L'Auto* (1900), were born to report on cycling, not on soccer. The bicycle was *the* dream for the

1 See John Carlin (2009).

poor. The hero was a skinny man, stooped on the handlebars, heaving his lungs on unpaved roads. But cycling, so rich in individual metaphor, lacked the social metaphor. The time was not about the individual but about the masses, for the flag. And cycling failed to express and capture the clan, the temple, the war, and eternity. All these phenomena, however, were in soccer: the clan (football club), the temple (the stadium), the war (the enemy is the other club or another city or another country), and eternity (a shirt and a flag whose—supposedly glorious—tradition is inherited by successive generations). In soccer, you never walk alone.

And so, more than a century after soccer was officially born, we find that 5% of all the people who *ever* lived on this planet watched the final of the last World Cup between Holland and Spain. Equivalently, if the idea of counting people not alive seems strange, sources report that about 50% of humans who were alive on July 11, 2010, watched the final of the last World Cup.

This book is an economics book, and soccer is the common thread of all the chapters. But the nature of the relationship between economics and soccer studied here is not entirely obvious. Over recent decades, economics has extended across many fields and conquered areas previously considered to belong to sociology, law, political science, history, biology, and other sciences. Economics can also conquer the analysis of sports, including the beautiful game, and, indeed, research has shown that economics can say many things about soccer and other sports.[2] But this is not the motivating idea behind this book. This is not a book about what economics can do for soccer. The idea is precisely the opposite: It is about what soccer can do for economics.

In the 16th and 17th centuries, stones falling from towers in Pisa and Florence sparked fundamental insights for Galileo Galilei (1564–1642) in his tests of the theory of gravity. Sir Isaac Newton (1642–1727) followed the same concept in the 17th and 18th centuries but worked with data from apples falling from trees. Apples or stones contributed to providing the empirical evidence necessary to evaluate a number of theories of physics for the first time. Just as data involving stones and apples were useful in physics, data from soccer are useful for economics. Soccer replaces apples and stones, and economics replaces physics. Using data from soccer, this book attempts to obtain and present novel insights into human behavior. This difference is what distinguishes this book from other economics books and from other books on the study of sports.

2 For instance, there is much published research on the economics of baseball and basketball, as well as much research on the sports business and the sports labor market (e.g., profitability of clubs, attendance demand, unionization, competitive balance, and so on) in several sports.

Talking about distinctions, "what most distinguishes economics as a discipline from other disciplines is not its subject matter but its approach. . . . The economic approach is applicable to all human behavior" (Becker 1976). In other words, because it is an approach applicable to all forms of human behavior, any type of data about human activity is potentially useful to evaluate economic theories. In fact, sports are in many ways the perfect laboratory for testing economic theories for a number of reasons. There is an abundance of readily available data, the goals of the participants are often uncomplicated (score, win, enforce the rules), and the outcomes are extremely clear. The stakes are typically high, and the subjects are professionals with experience. And so, "if one of the attractions of spectator sports is to see occasionally universal aspects of the human struggle in stark and dramatic forms, their attraction to economists is to illustrate universal economic principles in interesting and tractable ways" (Rosen and Sanderson 2001).

Soccer is, of course, not the only sport that could do this. It just so happens that there are sufficient situations, cases, and settings suitable for research in or around soccer to make it worth studying and presenting in a book on what sports can do for economics. To the best of my knowledge, to date no other sport has been shown to provide the same wealth of research opportunities as soccer to do just this. But I predict that other sports will do so in the near future.

There was no straightforward way to organize the various chapters in this book, and I struggled with this challenge. In the end, I decided to first present five chapters about various aspects of game theory, then to include a middle chapter tackling a finance question, and then to finish with five chapters about broader economic issues. This organization is just a vague categorization because the line that divides game theory, economics, and finance is not always clear. Some chapters could also belong to different sections, and even the sections could have been defined differently and the chapters grouped in a different way.

Chapter 1 is perhaps the clearest example of what motivates this book and what it sets out to accomplish. Recall the movie *A Beautiful Mind*, Oscar recipient for Best Picture in 2001. It portrays the life and work of John F. Nash Jr., who received the Nobel Prize in Economics in 1994. Perhaps you would think that after a movie and a Nobel Prize, the theories of Mr. Nash must have been solidly established and empirically validated on countless occasions. Right? Well, not quite. A class of his theories deals with how people should behave in strategic situations that involve what are known as "mixed strategies," that is, choosing among various possible strategies when no single one is always the best when you face a rational opponent. This first chapter uses data from a specific play in soccer (a penalty kick) with professional players to provide the first complete test of a fundamental theorem in game theory: the minimax theorem.

Whereas the first chapter studies mixed strategies in the field (specifically, on the field that is the soccer pitch), the second chapter studies the same strategies with the same professional subjects playing the same game not in real life with a soccer ball but in a laboratory setting with cards. The idea is to evaluate the extent to which experimental lab data generalize to central empirical features of field data when the lab features are deliberately closely matched to the field features.

The third chapter studies the determinants of the behavior observed in the field and in the lab in the previous two chapters, and the lessons we can learn for the design of field experiments.

The fourth chapter presents a neuroeconomic study whose purpose is to map or locate mixed strategies in the brain. Although recent research in the area of neuroeconomics has made substantial progress mapping neural activity in the brain for different decision-making processes in humans, this is the first time that neural activity involving minimax decision-making is documented.

The fifth and last chapter in this first half of the book could have easily belonged to the second half. The only reason it is here is that it takes advantage of the same type of situation (a penalty kick) that is studied in previous chapters. This chapter describes a novel behavioral bias not previously documented in a competitive situation: psychological pressure in a dynamic competitive environment arising from the state of the competition. The data come from penalty shoot-outs, which are a sequence of penalty kicks, and again involve professional subjects.

The halftime chapter, chapter 6, concerns a fundamental finance question: Is the efficient-markets hypothesis correct? Do prices adjust to publicly available new information rapidly and in an unbiased fashion, such that no excess returns can be earned by trading on that information? How can this hypothesis be tested when time flows continuously and it is not possible to ascertain that no news occurs while people are trading?

Finally, the last five chapters are each mainly concerned with different specific questions in economics. Chapter 7 studies the effect of social forces as determinants of behavior, in particular the role of social pressure as a determinant of corruption. There is of course a wide-ranging theoretical literature on the mechanisms of corruption, especially monetary ones like bribes or promotions. Likewise, there is an important literature on the influences of social factors and environments on individual behavior. And yet little is known empirically about social forces as determinants of corruption. This chapter presents the first empirical evidence linking these two areas.

Chapter 8 is concerned with the implications of incentives. The idea is that strong incentives may often have dysfunctional consequences, which may occur not only in cases of individual incentive contracts but also in settings where individuals compete with each other and are rewarded on a relative performance basis. These incentives may be particularly damaging if agents can devote resources not only to productive activities but also to depressing each other's output. Unfortunately, although anecdotal accounts of "back-stabbing," bad-mouthing, and similar activities are easy to find, there does not exist any systematic work documenting such responses. This chapter fills this gap.

Chapters 9 and 10 are concerned with the rationality of fear and with the role of incentives in overcoming emotions, respectively. They present two simple empirical studies testing a specific model of emotions with data from an extreme form of social pressure and violence (hooliganism) and from a unique experiment in Argentina, respectively. Finally, chapter 11 provides the first market-based test of taste-based discrimination in a setting where none of the classical difficulties of the conventional approach to test for discrimination are present.

The questions studied in this book are not just "cute" questions of interest to people who like sports.[3] Neither are they concerned with the idea of illuminating the world of sports through economics. Instead, the objective is to make progress in the world of economics through sports, specifically through soccer. The economic questions studied are tested with sports data the same way that physicists study apples falling from trees, not because they are interested in fruits but because they are interested in theories that are important in physics.

I am not entirely sure what the right audience is for this book. The book is a combination of novel and existing research findings, together with popular science and some stories and anecdotes. Most chapters have a number of ingredients that may appeal to different audiences, and no chapter has all the ingredients that would appeal exclusively to a single audience. It should be of interest to general readers, undergraduate students at all levels (but especially upper level), faculty and doctoral and postdoctoral researchers, practitioners, and certainly to

3 General audiences that like soccer and sports may enjoy the novel insights and approaches in *The Numbers Game* (2013) by Chris Anderson and David Sally, *Soccernomics* (2012) by Simon Kuper and Stefan Szymanski, and *Scorecasting* (2011) by Tobias J. Moscowitz and L. Jon Wertheim. These are excellent books and do not require a reader to have any type of technical sophistication. An introduction to the economics of sports with a wealth of references is *Playbooks and Checkbooks* by Stefan Szymanski (2009). *The Economics of Football* (2011) by Stephen Dobson and John Goddard, provides much more than an introduction to many aspects of the economics of soccer.

sports people and sports economists. And I hope it pays dividends for lay readers, other scholars, and even public policy decision makers.

From a teaching perspective, I believe that the book can be used as both an undergraduate and a graduate text to teach economics through sports. Complemented with readings about theory and evidence from other sports, it is possible to teach economics and the applications of economics in a number of important areas and topics in microeconomics (e.g., discrimination, incentives, and human preferences), finance, game theory, experimental economics, behavioral economics, and even certain topics in statistics and econometrics.

Although most chapters have a research orientation and are structured like research articles, I have attempted to write them in an accessible, and I hope eminently readable manner. Most readers of this book do not need a lecture about the importance of soccer in the world, the importance of economics, or the importance of measuring human behavior. Likewise, I felt that they would not need an exhaustive list of references and a complete biography in every chapter. Some might have welcomed additional readings, and perhaps more theoretical, statistical, and econometric details, but these days it is relatively easy to identify good survey articles on pretty much any topic or to get acquainted with the mathematical or econometric knowledge needed to grasp even the tiniest details of a specific methodology behind some results. With this notion in mind, I have purposely minimized the number of references in most chapters, hoping that by interrupting the reading as little as possible, readers may keep up their momentum and get more juices flowing.

The reader who is not a scholar, who will not become one, and who is interested in catching up or expanding his or her knowledge may begin with the references that are provided in the text. These references include some surveys and introductory books written in a variety of styles and levels of mathematical sophistication and directed at lay people as well as scholars of economics, game theory, finance, and other areas.

My main hope, besides presenting a few novel research findings, is to inspire some laypeople, students, scholars, and even policy makers to think about economics from a different and new perspective. I understand that my hope is ambitious.

A caveat for the true soccer and sports fan: I am not hoping that this book will change the way you watch the beautiful game or the way you think about your favorite team or players, or a number of situations on and off the pitch. Nevertheless, I warn you that this may happen. Be aware that this is a risk that you are taking with this book and that the effects on the role that soccer and other sports may play in your life after reading it may not be positive.

# FIRST HALF

# 1

# PELÉ MEETS JOHN VON NEUMANN IN THE PENALTY AREA

© TAWNG/STOCKFRESH

I thought there was nothing worth publishing until the Minimax Theorem was proved. As far as I can see, there could be no theory of games without that theorem.

—John von Neumann 1953

Much real-world strategic interaction cannot be fully understood with current tools. To make further progress, the field needs to gain more experience in applications to the real world.

—Game Theory Society 2006

The Hungarian national soccer team of the 1950s was one of the greatest soccer teams in the history of the 20th century. It played against England at Empire Wembley Stadium on November 25, 1953, in front of 105,000 people, in what was termed "the match of the century." Hungary was the world's number one ranked team and on a run of 24 unbeaten games. England was the world's number three ranked team, and unbeaten at Wembley for 90 years against teams from outside the

British Isles. In what was then considered a shocking result, Hungary beat England 6–3.

As a preparation for the 1954 World Cup in Switzerland, on May 23, 1954, England visited Budapest in the hope of avenging the 6–3 defeat. Instead, Hungary gave another master class, beating England 7–1. This score still ranks as England's worst defeat.

In those years, soccer was already the world's most popular sport, and Hungary was the best soccer team in the world—often considered one of the four or five best teams in history. As Olympic champion in 1952, not surprisingly, Hungary was the favorite to win the upcoming World Cup. Consistent with these expectations, Hungary easily beat Korea 9–0 and West Germany 8–3 in the first round. Then, it beat Brazil 4–2 in the quarterfinals and Uruguay, which had never been beaten in World Cup games, 4–2 in the semifinals. Its opponent in the final was West Germany, which surprisingly had managed to win all of its games after its initial defeat in the first round to Hungary. In the Wankdorf Stadium in Bern, 60,000 people saw Germany beat Hungary 3–2 in what was called "the Miracle of Bern." The sports announcer shouted in the background of the final scene of Rainer Werner Fassbinder's film *The Marriage of Maria Braun,* featuring this event, "Deutschland ist wieder was!" (Germany is something again!). This victory represented a powerful symbol of Germany's recovery from the ravages of the Second World War.

It is probably safe to assume that there were few Hungarians in the world in the 1950s who were not proudly aware of the accomplishments of their national team in the world's most popular sport. Indeed, Neumann Janos, born on December 28, 1903, in Budapest, a superb scientist who had migrated to the United States and was then known as John von Neumann, could not have been completely ignorant of the team's success.

John von Neumann is considered a scientific genius of the 20th century. A brilliant mathematician and physicist, he left a profound mark, with fundamental contributions in theoretical physics, applied physics, decision theory, meteorology, biology, economics, and nuclear deterrence; he became, more than any other individual, the creator of the modern digital computer.

"He was a genius, the fastest mind I have ever encountered. . . . He darted briefly into our domain, and it has never been the same since." Paul Samuelson (Nobel laureate in Economics, 1970, quoted in Macrae 1992) is referring here to the three fundamental contributions von Neumann made in economics: first, his 1928 paper "Zur Theorie der Gesellschaftsspiele," published in *Mathematische Annalen*, which established von Neumann as the father of game theory; second, his 1937 paper "A Model of General Equilibrium" (translated and published in 1945–46 in the *Review of Economic Studies*); and third, his classic book

*Theory of Games and Economic Behavior*, coauthored with Oskar Morgenstern in 1944 (Macrae 1992).

As a mathematician, von Neumann's own philosophical views induced him to choose to work in a variety of fields, and his selection of research questions and the resulting measure of his success were largely influenced by aesthetic values (von Neumann 1947). However, he also warned that mathematics loses much of its creative drive when too far removed from empirical sources. And yet, despite the place in the world of soccer that Hungary occupied, and despite soccer's place as the world's most popular game, everything indicates that he was not particularly interested in sports as an empirical source, or in the empirical verification of game theory theorems with sports data or with any data:

> The truth is that, to the best of my knowledge, my father had absolutely no interest in soccer or any other team sport, even as a spectator or news-follower. Ironically, he wasn't much on games in general (though he loved children's toys, which he could often persuade to yield up some principle of mathematics or physics); but his game-playing didn't go much beyond an occasional game of Chinese checkers at my request. I don't believe he even played poker.
>
> Warmest regards, Marina von Neumann Whitman
> (Private email correspondence, October 13, 2010)

As Kreps (1991) notes, "the point of game theory is to help economists understand and predict what will happen in economic, social and political contexts."[1] But if von Neumann considered, as the initial quotation suggests, that there could be no theory of games without proving the minimax theorem, then it seems appropriate to think that he would have considered that there could be no empirical applicability of the theory of games without first having verified empirically that theorem. As noted below, the minimax theorem was not empirically verified until 2003.

The empirical verification of strategic models of behavior is often difficult and problematic. In fact, testing the implications of *any* game theoretical model in a real-life setting has proven extremely difficult in the economics literature for a number of reasons. The primary reason is that many predictions often hinge on properties of the utility functions and the values of the rewards used. Furthermore, even when predictions are invariant over classes of preferences, data on rewards are seldom available in natural settings. Moreover, there is often great difficulty in determining the actual strategies available to the individuals involved, as well as in measuring these individuals' choices, effort levels, and the

1 Not everyone agrees that this is the point. See Rubinstein (2012).

incentive structures they face. As a result, even the most fundamental predictions of game-theoretical models have not yet been supported empirically in real situations.

Von Neumann published the minimax theorem in his 1928 article "Zur Theorie der Gesellschaftsspiele." The theorem essentially says,

> For every two-person, zero-sum game with finitely many strategies, there exists a value $V$ and a mixed strategy for each player, such that:
>
> (a) Given player 2's strategy, the best payoff possible for player 1 is $V$, and
> (b) Given player 1's strategy, the best payoff possible for player 2 is $-V$.
>
> Equivalently, Player 1's strategy guarantees him a payoff of $V$ regardless of Player 2's strategy, and similarly Player 2 can guarantee himself a payoff of $-V$.

A mixed strategy is a strategy consisting of possible moves and a probability distribution (collection of weights) that corresponds to how frequently each move is to be played. Interestingly, there are a number of interpretations of mixed strategy equilibrium, and economists often disagree as to which one is the most appropriate. See, for example, the interesting discussion in the classic graduate textbook by Martin Osborne and Ariel Rubinstein, *A Course on Game Theory* (1994, Section 3.2).

A game is called zero-sum or, more generally, constant-sum, if the two players' payoffs always sum to a constant, the idea being that the payoff of one player is always exactly the negative of that of the other player. The name "minimax" arises because each player minimizes the maximum payoff possible for the other. Since the game is zero-sum, he or she also minimizes his or her own maximum loss (i.e., maximizes his or her minimum payoff).

Most games or strategic situations in reality involve a mixture of conflict and common interest. Sometimes everyone wins, such as when players engage in voluntary trade for mutual benefit. In other situations, everyone can lose, as the well-known prisoner's dilemma situations illustrate. Thus, the case of *pure conflict* (or zero-sum or constant-sum or strictly competitive) games represents the extreme case of conflict situations that involve no common interest. As such, and as Aumann (1987) puts it, zero-sum games are a "vital cornerstone of game theory." It is not a surprise that they were the first to be studied theoretically.

The minimax theorem can be regarded as a special case of the more general theory of Nash (1950, 1951). It applies only to two-person, zero-sum or constant-sum games, whereas the Nash equilibrium concept can be used with any number of players and any mixture of conflict and common interest in the game.

Before undertaking a formal analysis, let us take a brief detour and look at the following play in soccer: a penalty kick. A penalty kick is awarded against a team that commits one of the 10 punishable offenses inside its own penalty area while the ball is in play. The world governing body of soccer, the Fédération Internationale de Football Association (FIFA), describes the simple rules that govern this play in the official *Laws of the Game* (FIFA 2012). First, the position of the ball and the players are determined as follows:

- "The ball is placed on the penalty mark in the penalty area.
- The player taking the penalty kick is properly identified.
- The defending goalkeeper remains on the goal line, facing the kicker, between the goalposts, until the ball has been kicked.
- The players other than the kicker are located inside the field of play, outside the penalty area, behind the penalty mark, and at least 10 yards (9.15 meters) from the penalty mark."

Then,

- "The player taking the penalty kicks the ball forward.
- He does not play the ball a second time until it has touched another player.
- A goal may be scored directly from a penalty kick."

The credit of inventing this play belongs to William McCrum. McCrum was a wealthy linen manufacturer, raconteur, cricketer, and the goalkeeper of Milford Everton, a small club in County Armagh, which played the inaugural season of the Irish Championship in 1890–91. History does not fully record how good a keeper he was, but he was certainly kept busy during that first Irish League season. Milford Everton finished at the bottom of the league with no points, a record of 10 goals scored, and 62 conceded, and the team was promptly relegated. McCrum may not have been one of the world's greatest goalkeepers, but he was a gentleman and justly proud of his reputation for good sportsmanship. His obituary in 1932 paints a picture of a man of honor who was frustrated and angry at the "win-at-all-costs" attitude that was poisoning his beloved soccer (Miller 1998).

McCrum believed that anyone who failed to abide by the spirit of the game should face a sanction that would punish not just the individual offender but also the whole team. Holding an influential position in the Irish Football Association, he submitted his proposal for a "penalty kick" to the association in 1890. Jack Reid, general secretary of the association, then formally forwarded McCrum's proposal to the international board for consideration at the board meeting to be held on June 2, 1890, and, he hoped, its subsequent incorporation into the laws. It immediately ran into a storm of protest. The reception was ferocious.

Press, administrators, and players publicly derided the idea. Some commentators even nicknamed the proposal the "death penalty," implying that it would be the death of the game as they knew it. Many people did not want to introduce a rule that effectively conceded that teams and players often resorted to unsporting methods. It was in this atmosphere that the Irish Football Association decided to withdraw the proposal. The international board, however, agreed to discuss the issue at the next meeting one year later. On June 2, 1891, somewhat surprisingly, the atmosphere was quite different and the proposal passed unanimously.

The penalty kick was born, albeit not in the form that we know it today. The new law came into force immediately, and, to be fair, it was not a huge success. There were obvious flaws in the first draft, and players—particularly goalkeepers—were quick to take advantage. Furthermore, gentlemen did not commit fouls. It took almost 40 years, until 1929, before the penalty law finally became what William McCrum intended it to be—an effective punishment for foul play. He lived to see his idea reach fruition but then died, a year later, after a long illness. (Trivia alert: On September 14, 1891, the Wolverhampton Wanderers were awarded the first penalty kick in a football league in a game against Accrington Stanley. The penalty was taken and scored by "Billy" Heath as the Wolves went on to win the game 5–0.)

McCrum's legacy is enormous, considering the worldwide importance of soccer today and the significance of the penalty kick within the game. He would have, no doubt, been proud to see how central his idea became to the overall development of the game. However, not even in his wildest dreams could he have anticipated that his penalty kick could also provide the data necessary to verify for the first time a mathematical theorem that was fundamental in economics: the minimax theorem. This is the objective of this chapter.

A formal model of the penalty kick may be written as follows. Let the player's payoffs be the probabilities of success ("score" for the kicker and "no score" for the goalkeeper) in the penalty kick. The kicker wishes to maximize the expected probability of scoring, and the goalkeeper wishes to minimize it. Consider, for example, a simple $2 \times 2$ game-theoretical model of player's actions for the penalty kick and let $\pi_{ij}$ denote the kicker's probabilities of scoring, where $i = \{L,R\}$ denotes the kicker's choice and $j = \{L,R\}$ the goalkeeper's choice, with $L$ = left, $R$ = right:

| | $L$ | $R$ |
|---|---|---|
| $L$ | $\pi_{LL}$ | $\pi_{LR}$ |
| $R$ | $\pi_{RL}$ | $\pi_{RR}$ |

Each penalty kick involves two players: a kicker and a goalkeeper. In the typical kick in professional leagues, the ball takes about 0.3 seconds to travel the distance between the penalty mark and the goal line. This is less time than it takes for the goalkeeper to react and move to the possible paths of the ball. Hence, both kicker and goalkeeper must move simultaneously. Players have few strategies available, and their actions are observable. There are no second penalties in the event that a goal is not scored. The initial locations of both the ball and the goalkeeper are always the same: The ball is placed on the penalty mark, and the goalkeeper positions himself on the goal line, equidistant from the goalposts. The outcome is decided, in effect, immediately (roughly within 0.3 seconds) after players choose their strategies.

The clarity of the rules and the detailed structure of this simultaneous one-shot play capture the theoretical setting of a zero-sum game extremely well. In this sense, it presents notable advantages over other plays in professional sports and other real-world settings. In baseball, pitchers and batters have many actions available, and there are numerous possible outcomes. In cricket and tennis, possible outcomes are limited, but players also have many strategic choices available. Even in these sports, the direction of the serve or the pitch, its spin, and the initial location of the opponent are all important strategic choices that are hard to quantify. For instance, the position of the player returning a tennis serve or attempting to hit a baseball affects the choice of strategy by the server or the pitcher. A key additional difficulty is that a serve or a pitch is not a simultaneous (static) but a sequential (dynamic) game, in that the outcome of the play is typically not decided immediately. After a player serves or a pitcher throws, often there is subsequent strategic play that plays a crucial role in determining the final outcome. Each point in these situations is more like part of a dynamic game with learning, where each player plays what in economics is known as a multi-armed bandit problem at the start of the match.[2] As such, these situations deviate substantially from the theoretical postulates put forward here, and notable difficulties arise both in modeling nonsimultaneous situations theoretically and in observing all strategic choices in a given play.

The penalty kick game has a unique Nash equilibrium in mixed strategies when

2 In a dynamic game, there probably are spillovers from point to point, whereas in a standard repeated zero-sum game, especially if repeated infrequently, there are no such payoff spillovers. For instance, in tennis, having served to the left on the first serve (and say, faulted) may effectively be "practice" in a way that makes the server momentarily better than average at serving to the left again. If this effect is important, the probability that the next serve should be inside the line should increase. In other words, there should be negative serial correlation in the choice of serve strategies rather than, as will be shown later, the random play (no serial correlation) that is predicted by minimax. Consistent with this hypothesis, the results in Walker and Wooders (2001) confirm that tennis players switch serving strategies too often to be consistent with random play.

$$\pi_{LR} > \pi_{LL} < \pi_{RL},$$
$$\pi_{RL} > \pi_{RR} < \pi_{LR}$$

If the play in a penalty kick can be represented by this model, then equilibrium play requires each player to use a mixed strategy. In this case, the equilibrium yields two sharp testable predictions about the behavior of kickers and goalkeepers:

1. Success probabilities—the probability that a goal will be scored (not scored) for the kicker (goalkeeper)—should be the same across strategies for each player.

   Formally, let $g_L$ denote the goalkeeper's probability of choosing left. This probability should be chosen so as to make the kicker's probability of success identical across strategies. That is, $g_L$ should satisfy $pk_L = pk_R$, where

   $$pk_L = g_L \cdot \pi_{LL} + (1 - g_L) \cdot \pi_{LR}$$
   $$pk_R = g_L \cdot \pi_{RL} + (1 - g_L) \cdot \pi_{RR}$$

   Similarly, the kicker's probability of choosing left, $k_L$, should be chosen so as to make the goalkeeper's success probabilities identical across strategies, $pg_L = pg_R$, where

   $$pg_L = k_L \cdot (1 - \pi_{LL}) + (1 - k_L) \cdot (1 - \pi_{RL})$$
   $$pg_R = k_L \cdot (1 - \pi_{LR}) + (1 - k_L) \cdot (1 - \pi_{RR})$$

2. Each player's choices must be serially independent given constant payoffs across games (penalty kicks). That is, individuals must be concerned only with instantaneous payoffs, and intertemporal links between penalty kicks must be absent. Hence, players' choices must be independent draws from a random process. Therefore, they should not depend on one's own previous play, on the opponent's previous play, on their interaction, or on any other previous actions.

The intuition for these two testable hypotheses is the following. In a game of pure conflict (zero-sum), if it would be disadvantageous for you to let your opponent see your actual choice in advance, then you benefit by choosing at random from your available pure strategies. The proportions in your mix should be such that the opponent cannot exploit your choice by pursuing any particular pure strategy out of those available to him or her—that is, each player should get the same average payoff when he or she plays any of his or her pure strategies against his or her opponent's mixture.

In what follows, we test whether these two hypotheses can be rejected using classical hypothesis testing and real data. Incidentally, this reject–no reject dichotomy may be quite rigid in situations where the theory makes precise point predictions, as in the zero-sum game that we study.[3]

Data were collected on 9,017 penalty kicks during the period September 1995–June 2012 from professional games in Spain, Italy, England, and other countries. The data come from a number of TV programs, such as the *English Soccer League* in the United States, *Estudio Estadio* and *Canal+ Fútbol* in Spain, *Novantesimo Minuto* in Italy, *Sky Sports Football* in the United Kingdom, and others. These programs review all of the best plays in the professional games played every week, including all penalty kicks that take place in each game. The data include the names of the teams involved in the match, the date of the match, the names of the kicker and the goalkeeper for each penalty kick, the choices they take (left, center, or right), the time at which the penalty kick is shot, the score at that time, and the final score in the match. They also include the kicker's kicking leg (left or right) and the outcome of the shot (goal or no goal).[4] Around 80% of all observations come from league matches in England, Spain, and Italy.[5] Together with Germany, these leagues are considered to be the most important in the world.

There are two types of kickers, depending on their kicking legs: left-footed and right-footed. Most kickers in the sample are right-footed, as is the case in the population of soccer players and in the general population. These two types have different strong sides. Left-footed kickers shoot more often to the left-hand side of the goalkeeper than to the right-hand side, whereas right-footed kickers shoot more often to the right-hand side. Basic anatomical reasons explain these different strengths.

To deal with this difference, it makes sense to "normalize" the game and rename choices according to their "natural sides." In other words, given that the roles are reversed for right-footed kickers and left-footed kickers, it would be erroneous to treat the games associated with these different types of kickers as equal. For this reason, in the remainder of the chapter, players' choices are renamed according to the kickers' natural sides. Whatever the kicker's strong foot actually is, $R$ denotes "kicker's natural side" and $L$ denotes "kicker's nonnatural side." When the kicker is right-footed, the natural side $R$ is the right-hand side of the goalkeeper, and when the kicker is left-footed, it is the left-hand side of the goalkeeper. This notation means, for instance, that a left-footed

3 O'Neill (1991) suggests for these cases an alternative that is much less rigid than the reject–no reject dichotomy: a Bayesian approach to hypothesis testing combined with a measure of closeness of the results to the predictions.

4 The outcome "no goal" includes saves made by the goalkeeper and penalties shot wide, to the goalpost, or to the crossbar by the kicker, each in separate categories.

5 The rest come from cup competitions (elimination tournaments that are simultaneously played during the annual leagues) and from international games.

kicker kicking to the goalkeeper's right is the same as a right-footed kicker kicking to the goalkeeper's left. Thus, the goalkeeper plays the same game when he or she faces a left-footed or a right-footed kicker, but the actions are simply identified differently. All that matters is whether the kicker and goalkeeper pick the kicker's strong side $R$ or the kicker's weak side $L$. Payoffs are the same for the two kicker types up to the renaming of the actions. The same argument goes for goalkeepers. They tend to choose right more often than left when facing a right-footed kicker and left more often than right when facing a left-footed kicker, but the scoring rates are statistically identical when they face the two player types after the renaming of the actions.[6]

Table 1.1 shows the relative proportions of the different choices made by the kicker and the goalkeeper (Left ($L$), Center ($C$), or Right ($R$)), with the total number of observations in the second left-most column. The first letter refers to the choice made by the kicker and the second to the choice made by the goalkeeper, both from the viewpoint of the goalkeeper. For instance, "*RL*" means that the kicker chooses to kick to the right-hand side of the goalkeeper and the goalkeeper chooses to jump to his or her left. The right-most column shows the scoring rate for a given score difference. The term "score difference" is defined as the number of goals scored by the kicker's team minus the number of goals scored by the goalkeeper's team at the time the penalty is shot. For instance, a −1 means that the kicker's team was behind by one goal at the time of the penalty kick.

The strategy chosen by goalkeepers coincides with the strategy followed by kickers in about half of all penalty kicks in the data set. Most are *RR* (30.5%); 16.7% are *LL*, and 0.9% are *CC*. Kickers kick to the center relatively rarely (6.8% of all kicks), whereas goalkeepers appear to choose $C$ even less often (3.5%), perhaps because they already cover part of the center with their legs when they choose $R$ or $L$. The percentage of kicks where players' strategies do not coincide with each other is almost equally divided between *LR* (21.6%) and *RL* (21.7%).

A goal is scored in 80.07% of all penalty kicks. The scoring rate is close to 100% when the goalkeeper's choice does not coincide with the kicker's, and it is over 60% when it coincides. The average number of goals per match in the sample is 2.59. It is thus no surprise to observe that in most penalty kicks the score difference is 0, 1, or −1 at the time of the shot. For these score differences, the scoring rate is slightly greater in

6 See Palacios-Huerta (2003). This statistical identity can be shown using a regression framework. The null hypothesis that kicker's types are perfectly symmetric corresponds to a finding that kicker-type fixed effects are jointly insignificantly different from zero in explaining whether a goal was scored, including variables that describe the state of the soccer match at the time the penalty is shot as controls. The same holds for goalkeepers facing the different types.

**Table 1.1.** Distribution of Strategies and Scoring Rates in Percentage Terms

| Score Difference | #Obs | *LL* | *LC* | *LR* | *CL* | *CC* | *CR* | *RL* | *RC* | *RR* | Scoring Rate |
|---|---|---|---|---|---|---|---|---|---|---|---|
| 0 | 3701 | 16.5 | 1.2 | 21.5 | 4.0 | 1.4 | 3.5 | 20.6 | 1.2 | 30.1 | 81.5 |
| 1 | 1523 | 15.1 | 0.5 | 16.2 | 4.2 | 1.5 | 2.6 | 29.9 | 0.5 | 29.5 | 78.1 |
| -1 | 2001 | 16.1 | 1.5 | 23.3 | 2.1 | 0.0 | 1.7 | 20.3 | 1.0 | 34.0 | 80.3 |
| 2 | 607 | 11.9 | 3.4 | 19.4 | 5.2 | 1.5 | 0.7 | 23.7 | 1.7 | 32.5 | 75.7 |
| -2 | 744 | 20.1 | 1.5 | 27.6 | 3.7 | 0.0 | 2.5 | 16.1 | 0.3 | 28.2 | 78.5 |
| Others | 441 | 28.8 | 0.5 | 27.3 | 0.5 | 0.6 | 1.4 | 17.7 | 0.5 | 22.7 | 82.4 |
| All | 9017 | 16.7 | 1.7 | 21.6 | 3.4 | 0.9 | 2.5 | 21.7 | 0.9 | 30.5 | 80.07 |

tied matches (81.5%), followed by the rate in matches where the kicker's team is behind by one goal (80.3%), and then by the rate in matches where his or her team is ahead by one goal (78.1%).

Before we begin any formal test, it is worth examining the extent to which observed behavior appears to be close to the Nash equilibrium predictions. Players in the sample choose either $R$ or $L$ 96.3% of the time, kickers 93.2% of the time, and goalkeepers, 96.5%.[7] In what follows, we consider the choice $C$ as if it was the same as the natural choices.[8] The typical penalty kick may then be described by the simple $2 \times 2$ model outlined earlier. Thus a penalty kick has a *unique* Nash equilibrium, and the equilibrium requires each player to use a mixed strategy. As mentioned already, equilibrium theory makes two testable predictions about the behavior of kickers and goalkeepers: (1) Winning probabilities should be the same across strategies for both players, and (2) each player's strategic choices must be serially independent.

For all players in the sample, the empirical scoring probabilities are the following:

| | $g_L$ | $1-g_L$ |
|---|---|---|
| $k_L$ | 59.11 | 94.10 |
| $1-k_L$ | 93.10 | 71.22 |

where, as indicated above, $k_L$ and $g_L$ denote the nonnatural sides. We can now compute the mixed strategy Nash equilibrium in this game (minimax frequencies) and compare it with the actual mixing probabilities observed in the sample (see figures 1.1 and 1.2). Interestingly, we find that observed aggregate behavior is virtually *identical* to the theoretical predictions:

| | $g_L$ | $1-g_L$ | $k_L$ | $1-k_L$ |
|---|---|---|---|---|
| Nash Predicted Frequencies | 40.23% | 59.77% | 38.47% | 61.53% |
| Actual Frequencies | 41.17% | 58.83% | 38.97% | 61.03% |

7 Chiappori et al. (2002) study the aggregate predictions of a zero-sum game, rather than individual player choices and pay close attention to the possibility that $C$ is an available pure strategy. They conclude that the availability of $C$ as an action is not an issue. Their findings are also substantiated in the data set used in this chapter. This evidence means that a penalty kick may be described as a two-action game.

8 Professional players basically consider strategy $C$ and the strategy of playing their natural side as equivalent. The reason is that they typically kick with the interior side of their foot, which allows for greater control of the kick, by approaching the ball running from their nonnatural side. This phenomenon makes choosing $C$ or their natural side equally difficult.

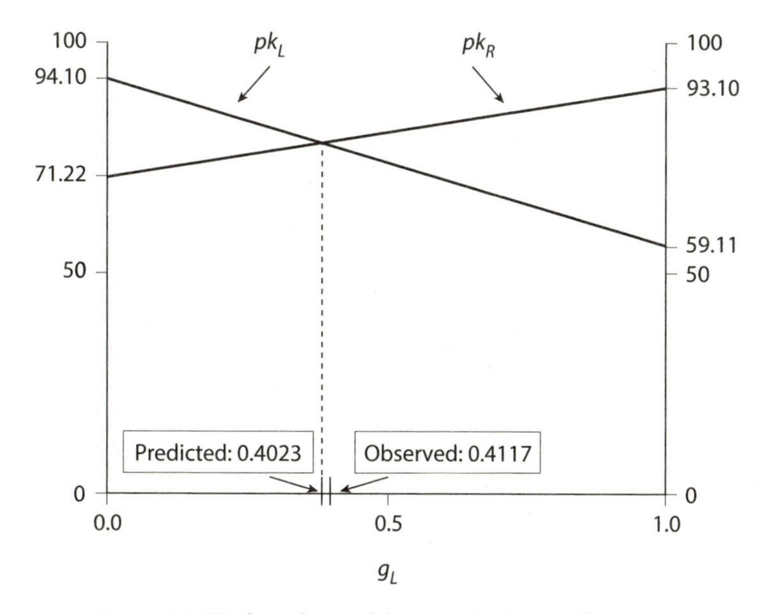


**Figure 1.1.** Nash and actual frequencies for goalkeepers.

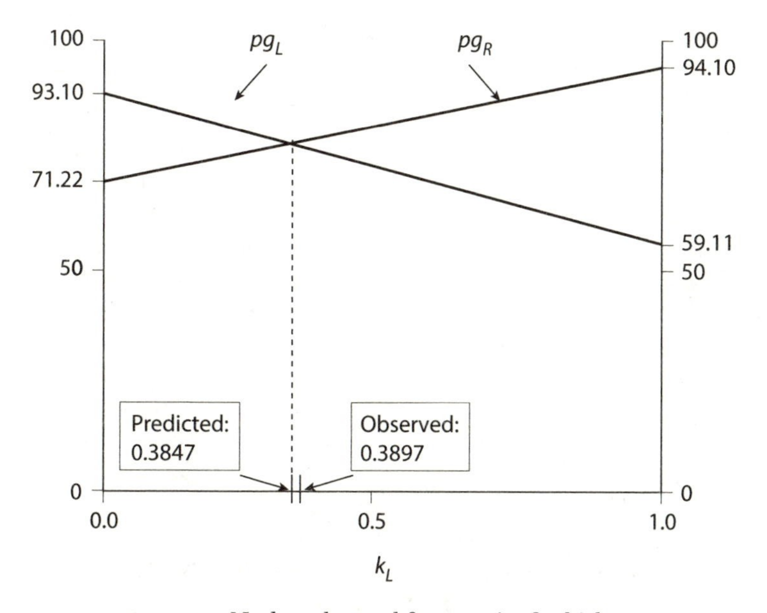


**Figure 1.2.** Nash and actual frequencies for kickers.

This is, at the very least, encouraging for the model. We turn next to testing the implications of the minimax theorem.

## IMPLICATION NUMBER 1: TESTS OF EQUAL SCORING PROBABILITIES

The tests of the null hypothesis that the scoring probabilities for a player (kicker or goalkeeper) are identical across strategies can be implemented with the standard proportions tests, that is, using Pearson's $\chi^2$ goodness-of-fit test of equality of two distributions.

Let $p_{ij}$ denote the probability that player $i$ will be successful when choosing strategy $j \in \{L,R\}$, $n_{ij}$ the number of times that $i$ chooses $j$, and $N_{ijS}$ and $N_{ijF}$ the number of times in which player $i$ chooses strategy $j$ and is successful ($S$) or fails ($F$) in the penalty kick. Success for a kicker is to score a goal, and for a goalkeeper is that a goal is not scored. Hence, we want to test the null hypothesis $p_{iL} = p_{iR} = p_i$. Statisticians tell us that to do this, the Pearson statistic for player $i$

$$p^i = \sum\nolimits_{j \in \{L,R\}} \left[ \frac{(N_{ijS} - n_{ij} p_i)^2}{n_{ij} p_i} + \frac{(N_{ijF} - n_{ij}(1 - p_i))^2}{n_{ij}(1 - p_i)} \right]$$

is distributed asymptotically as a $\chi^2$ with 1 degree of freedom.

Quick statistical detour. In this and other statistical tests in this book, we will report $p$-values, so it is important to have a sense of what they are. Under the assumption that the hypothesis of interest (called the null hypothesis) is true, the $p$-value is the probability of obtaining a test statistic at least as extreme as the one that is actually observed. Thus, one often "rejects the null hypothesis" when the $p$-value is less than a predetermined significance level, often 0.05 (5%) or 0.01 (1%), indicating that the observed result would be highly unlikely under the null hypothesis. Many common statistical tests in this book, such as $\chi^2$ tests or Student's $t$-test, produce test statistics that will be interpreted using $p$-values.

It is also possible to study whether behavior at the aggregate level is consistent with equilibrium play by testing the joint hypothesis that each individual case is *simultaneously* generated by equilibrium play. The test statistic for the Pearson joint test in this case is the sum of all the $N$ individual test statistics, and under the null hypothesis this test is distributed as a $\chi^2$ with $N$ degrees of freedom. Note that this joint test allows for differences in probabilities $p_i$ across players.

## IMPLICATION NUMBER 2: TESTS OF RANDOMNESS OR SERIAL INDEPENDENCE

The second testable implication is that a player's mixed strategy is the same at each penalty kick. This notion implies that players' strategies are random or serially independent. Their play is not serially independent if, for instance, they choose not to switch their actions often enough or if they switch actions too often.

The work on randomization is extensive in the experimental economics and psychological literatures. Interestingly, this hypothesis has never found support in any empirical (natural and experimental) tests of the minimax hypothesis, and it is rarely supported in other tests. In particular, when subjects are asked to generate random sequences, their sequences typically have negative autocorrelation, that is, individuals exhibit a bias *against* repeating the same choice.[9] This phenomenon is often referred to as the "law of small numbers" (subjects may try to reproduce, in short sequences, what they know are the properties of long sequences). The only possible exception is Neuringer (1986), who explicitly taught subjects to choose randomly after hours of training by providing them with detailed feedback from previous blocks of responses in an experiment. These training data are interesting in that they suggest that experienced subjects might be able to learn to generate randomness. As Camerer (1995) remarks, "whether they do in other settings, under natural conditions, is an empirical question."

A simple way to test for randomness is to use the standard "runs test." Consider the sequence of strategies chosen by a player in the order in which they occurred $s = \{s_1, s_2, \ldots, s_n\}$ where $s_x \in \{L,R\}$, $x \in [1,n]$, and $n = n_R + n_L$ are the number of natural side and nonnatural side choices made by the player. Let $r$ denote the total number of runs in the sequence $s$. A run is defined as a succession of one or more identical symbols that are followed and preceded by a different symbol or no symbol at all. Let $f(r, s)$ denote the probability that there are exactly $r$ runs in the sequence $s$. Let $\Phi[r, s] = \Sigma_{k=1,\ldots,r} f(k,s)$ denote the probability of obtaining $r$ or fewer runs. Gibbons and Chakraborti (1992) show that by using the exact mean and variance of the number of runs in an ordered sequence, then, under the null hypothesis that strategies are serially independent, the critical values for the rejection of the hypothesis can be found from the Normal distribution approximation to the null distribution.

More precisely, the variable

$$\frac{r - 0.5 - 2\left(\frac{n_L n_R}{n}\right)}{\left[2n_L n_R\left(\frac{2n_L n_R - n}{n^2(n-1)}\right)\right]^2}$$

9 See Bar-Hillel and Wagenaar (1991), Rapoport and Budescu (1992), Rapoport and Boebel (1992), and Mookherjee and Sopher (1994). Neuringer (2002), Rabin (2002) and Camerer (1995) review the literature. See also Tversky and Kahneman (1971).

is distributed as a standardized Normal probability distribution $\mathcal{N}(0,1)$. The null hypothesis will then be rejected at the 5% confidence level if the probability of $r$ or fewer runs is less than 0.025 or if the probability of $r$ or more runs is less than 0.025, that is, if $\Phi[r, s] < 0.025$ or if $1 - \Phi[r-1, s] < 0.025$. Similarly, at the 10% level, the hypothesis is rejected if they are less than 0.05.[10]

The results in table 1.2 show the results of the Pearson test and the runs test for 40 world-class soccer players, half kickers, and half goalkeepers.

The null hypothesis of equality of payoffs cannot be rejected for the majority of players. It is rejected for just two players (David Villa and Frank Lampard) at the 5% significance level and four players at the 10% significance level (in addition to Villa and Lampard, Iker Casillas and Morgan De Sanctis). Note that we should expect some rejections, just as if we flip 40 coins 10 times each we should expect some coins, but not many, to yield by pure chance proportions that are far from 50–50, such as 9 heads and 1 tail, or 8 heads and 2 tails. The confidence levels we are willing to adopt (typically no greater than 5% or 10%) tell us how many rejections we should expect. In our case, with 40 players the expected number of rejections at the 5% level is $0.05 \times 40 = 2$, and at the 10% level, it is $0.10 \times 40 = 4$.

Thus, the evidence indicates that the hypothesis that scoring probabilities are identical across strategies cannot be rejected at the individual level for most players at conventional significance levels. The number of rejections is, in fact, identical to the theoretical predictions.

Furthermore, behavior at the aggregate level also appears to be consistent with equilibrium play. As already indicated, the joint hypothesis that each case is simultaneously generated by equilibrium play can be tested by computing the aggregate Pearson statistic (summing up the individual Pearson statistics) and checking if it is distributed as a $\chi^2$ with $\mathcal{N}$ degrees of freedom. The results show that the Pearson statistic is 36.535 and its associated $p$-value is 0.627 for all 40 players. Hence, the hypothesis of equality of winning probabilities cannot be rejected at the aggregate level. Focusing only on kickers, the relevant statistic is 20.96 with a $p$-value of 0.399, and for goalkeepers it is 15.58 with a $p$-value of 0.742. Hence, the hypothesis of equality of winning probabilities cannot be rejected for either subgroup.

With respect to the null hypothesis of randomness, the runs tests show that this hypothesis cannot be rejected for the majority of players. They neither appear to switch strategies too often or too infrequently,

10 Aggregate level tests may also be implemented by checking if the values in columns $\Phi[r, s]$ and $\Phi[r-1, s]$ tend to be uniformly distributed in the interval [0, 1], which is what should happen under the null hypothesis of randomization. See Palacios-Huerta (2003).

but just about the right amount. This hypothesis is in fact rejected for just three players (David Villa, Alvaro Negredo, and Edwin Van der Sar), and four players (in addition, Jens Lehman) at the 5% and 10% significance levels. For the same reasons as in the previous test, we should be expecting two and four rejections.

The runs test is simple and intuitive. However, it is a test that has low power to identify a lack of randomness. Put differently, current choices may be explained, at least in part, by past variables such as past choices or past outcomes, or past choices of the opponent, or interactions with these variables, and still the number of runs in the series of choices may appear to be neither too high nor too low. As such, many potential sources of dynamic dependence cannot be detected with a runs test. For this reason, some researchers on randomization have studied whether past choices or outcomes have any role in determining current choices by estimating a logit equation for each player. For instance, in Brown and Rosenthal (1990), the dependent variable is a dichotomous indicator of the current choice of strategy, and the independent variables are first and second lagged indicators for both players' past choices, the products of their first and second lagged choices, and an indicator for the opponent's current choices. The results show that in fact it is possible to detect a number of dynamic dependences with this logit equation that are not possible to detect with the runs test.[11]

Unfortunately, the standard logit equation is still problematic in that the way this procedure is typically implemented generates *biased* estimates. We will take a quick technical detour to explain why. The choice of strategy in a penalty kick may depend on certain observed characteristics of the player and his or her opponent, the specific sequence of past choices and past outcomes, and perhaps other variables. It may also depend on unobserved characteristics. Thus, the basic econometric problem is to estimate a binary choice model with lagged endogenous variables and unobserved heterogeneity where the effect of state dependence needs to be controlled for appropriately. The econometric estimation of these models is subject to a number of technical difficulties. For example, parameter estimates jointly estimated with individual fixed effects can be seriously *biased* and *inconsistent*. Arellano and Honoré (2001) offer an excellent review.

To establish the idea that past choices have no significant role in determining current choices, we estimate a logit equation for each

11 Compare table IV in Brown and Rosenthal (1990) with table 4 in Walker and Wooders (2001). There are many subjects that pass the runs test but still exhibit serial dependence in that a number of lagged endogenous variables (choices and outcomes) help predict their subsequent choices.

**Table 1.2.** Pearson and Runs Tests

| Name | #Obs | Proportions | | Success Rate | | Pearson Tests | | Runs Tests | | |
|---|---|---|---|---|---|---|---|---|---|---|
| | | $L$ | $R$ | $L$ | $R$ | Statistic | $p$-value | $r$ | $\Phi[r-1, s]$ | $\Phi[r, s]$ |
| Kickers: | | | | | | | | | | |
| Mikel Arteta | 53 | 0.433 | 0.566 | 0.782 | 0.833 | 0.218 | 0.639 | 27 | 0.439 | 0.551 |
| Alessandro Del Piero | 55 | 0.345 | 0.654 | 0.736 | 0.805 | 0.344 | 0.557 | 24 | 0.237 | 0.339 |
| Samuel Eto'o | 62 | 0.419 | 0.580 | 0.769 | 0.805 | 0.120 | 0.728 | 28 | 0.165 | 0.239 |
| Diego Forlán | 62 | 0.419 | 0.580 | 0.769 | 0.805 | 0.120 | 0.728 | 30 | 0.327 | 0.427 |
| Steven Gerrard | 50 | 0.340 | 0.660 | 0.823 | 0.909 | 0.777 | 0.377 | 23 | 0.382 | 0.507 |
| Thierry Henry | 44 | 0.477 | 0.522 | 0.809 | 0.782 | 0.048 | 0.825 | 19 | 0.086 | 0.145 |
| Robbie Keane | 42 | 0.309 | 0.690 | 0.769 | 0.758 | 1.174 | 0.278 | 17 | 0.184 | 0.296 |
| Frank Lampard | 38 | 0.236 | 0.763 | 0.666 | 0.793 | 4.113 | 0.042** | 17 | 0.791 | 0.898 |
| Lionel Messi | 45 | 0.377 | 0.622 | 1.000 | 0.928 | 1.270 | 0.259 | 22 | 0.416 | 0.544 |
| Alvaro Negredo | 45 | 0.288 | 0.711 | 0.769 | 0.906 | 1.501 | 0.220 | 26 | 0.986** | 0.995 |
| Martín Palermo | 55 | 0.381 | 0.618 | 0.714 | 0.735 | 0.028 | 0.865 | 23 | 0.098 | 0.158 |
| Andrea Pirlo | 39 | 0.384 | 0.615 | 0.733 | 0.833 | 0.566 | 0.451 | 20 | 0.505 | 0.639 |
| Xabi Prieto | 37 | 0.324 | 0.675 | 0.833 | 0.880 | 0.151 | 0.697 | 16 | 0.256 | 0.392 |
| Franck Ribéry | 38 | 0.500 | 0.500 | 0.789 | 0.736 | 0.145 | 0.702 | 25 | 0.930 | 0.964 |
| Ronaldinho | 46 | 0.456 | 0.543 | 0.952 | 0.880 | 0.753 | 0.385 | 24 | 0.460 | 0.580 |
| Cristiano Ronaldo | 51 | 0.372 | 0.627 | 0.842 | 0.718 | 1.008 | 0.315 | 24 | 0.342 | 0.458 |
| Roberto Soldado | 40 | 0.400 | 0.600 | 0.937 | 0.750 | 2.337 | 0.126 | 21 | 0.539 | 0.667 |
| Francesco Totti | 47 | 0.489 | 0.510 | 0.782 | 0.833 | 0.195 | 0.658 | 20 | 0.070 | 0.119 |
| David Villa | 52 | 0.365 | 0.634 | 0.631 | 0.909 | 5.978 | 0.014** | 18 | 0.010 | 0.022** |
| Zinedine Zidane | 61 | 0.377 | 0.622 | 0.782 | 0.815 | 0.099 | 0.752 | 26 | 0.126 | 0.192 |
| All | 962 | 0.386 | 0.613 | 0.795 | 0.822 | 20.96 | 0.399 | | | |

| | | | | | | | | | | |
|---|---|---|---|---|---|---|---|---|---|---|
| Goalkeepers: | | | | | | | | | | |
| Dani Aranzubia | 68 | 0.455 | 0.544 | 0.225 | 0.189 | 0.138 | 0.709 | 29 | 0.062 | 0.098 |
| Gianluigi Buffon | 71 | 0.408 | 0.591 | 0.241 | 0.142 | 1.113 | 0.291 | 35 | 0.420 | 0.518 |
| Willie Caballero | 60 | 0.350 | 0.650 | 0.095 | 0.230 | 1.674 | 0.195 | 29 | 0.522 | 0.634 |
| Iker Casillas | 69 | 0.347 | 0.652 | 0.250 | 0.088 | 3.278 | 0.070* | 32 | 0.414 | 0.520 |
| Petr Čech | 82 | 0.414 | 0.585 | 0.235 | 0.187 | 0.276 | 0.590 | 38 | 0.224 | 0.298 |
| Júlio César | 68 | 0.308 | 0.691 | 0.238 | 0.106 | 2.007 | 0.156 | 34 | 0.840 | 0.900 |
| Morgan De Sanctis | 62 | 0.435 | 0.564 | 0.148 | 0.342 | 3.018 | 0.082* | 34 | 0.700 | 0.783 |
| Tim Howard | 67 | 0.402 | 0.597 | 0.222 | 0.225 | 0.000 | 0.978 | 30 | 0.169 | 0.241 |
| Bodo Illgner | 68 | 0.352 | 0.647 | 0.250 | 0.272 | 0.041 | 0.839 | 33 | 0.547 | 0.650 |
| Gorka Iraizoz | 73 | 0.424 | 0.575 | 0.129 | 0.142 | 0.028 | 0.865 | 32 | 0.106 | 0.157 |
| David James | 69 | 0.391 | 0.608 | 0.185 | 0.238 | 0.270 | 0.603 | 40 | 0.924 | 0.954 |
| Oliver Kahn | 58 | 0.379 | 0.620 | 0.227 | 0.138 | 0.747 | 0.387 | 33 | 0.881 | 0.928 |
| Andreas Koepke | 70 | 0.428 | 0.571 | 0.233 | 0.150 | 0.787 | 0.374 | 31 | 0.119 | 0.175 |
| Jens Lehmann | 72 | 0.444 | 0.555 | 0.218 | 0.225 | 0.004 | 0.949 | 28 | 0.014 | 0.026* |
| Andrés Palop | 66 | 0.439 | 0.560 | 0.206 | 0.297 | 0.694 | 0.404 | 34 | 0.498 | 0.597 |
| Pepe Reina | 55 | 0.418 | 0.581 | 0.173 | 0.187 | 0.016 | 0.897 | 31 | 0.778 | 0.852 |
| Mark Schwarzer | 55 | 0.381 | 0.618 | 0.238 | 0.264 | 0.048 | 0.825 | 31 | 0.846 | 0.904 |
| Stefano Sorrentino | 48 | 0.458 | 0.541 | 0.136 | 0.269 | 1.275 | 0.258 | 27 | 0.687 | 0.783 |
| Víctor Valdes | 71 | 0.394 | 0.605 | 0.214 | 0.232 | 0.032 | 0.857 | 32 | 0.196 | 0.272 |
| Edwin van der Sar | 80 | 0.412 | 0.587 | 0.121 | 0.148 | 0.125 | 0.722 | 26 | 0.000 | 0.001** |
| All | 1332 | 0.402 | 0.597 | 0.199 | 0.198 | 15.58 | 0.742 | | | |

*Notes*: ** and * denote rejections at the 5% and 10% levels, respectively.

**Table 1.3.** Results of Significance Tests (Logit) for the Choice of the Natural Side

| Null Hypothesis | | Players Whose Behavior Allows Rejection of the Null Hypothesis at the: 5% Level | 10% Level |
|---|---|---|---|
| A. $a_1 = a_2 = b_0 = b_1 = b_2 = c_1 = c_2 = 0$ | Kicker | — | David Villa, Frank Lampard |
| | Goalkeeper | — | Iker Casillas |
| B. $a_1 = a_2 = 0$ | Kicker | — | David Villa |
| | Goalkeeper | — | Andreas Kopke |
| C. $b_1 = b_2 = 0$ | Kicker | — | — |
| | Goalkeeper | — | Jens Lehman |
| D. $c_1 = c_2 = 0$ | Kicker | — | Martín Palermo |
| | Goalkeeper | — | — |
| E. $b_0 = 0$ | Kicker | — | Ronaldinho |
| | Goalkeeper | — | Júlio César, Edwin van der Sar |

Here is the estimating equation:

$$R = G[a_0 + a_1\text{lag}(R) + a_2\text{lag2}(R) + b_0R^* + b_1\text{lag}(R^*) + b_2\text{lag2}(R^*) + c_1\text{lag}(R)\text{lag}(R^*) + c_2\text{lag2}(R)\text{lag2}(R^*)]$$

*Notes*: The asterisk * denotes the choice of the opponent. The terms "lag" and "lag2" denote the choices previously followed in the ordered sequence of penalty kicks. The function $G[x]$ denotes $\exp(x)/[1+\exp(x)]$.

player based on the Arellano and Carrasco (2003) method using the same specification as Brown and Rosenthal (1990). The model generates *unbiased* and *consistent* estimates and it allows for unobserved heterogeneity and for individual effects to be correlated with the explanatory variables (see table 1.3).

The main finding is that the null hypothesis of randomization (implication number 2), that all the explanatory variables are jointly statistically insignificant (hypothesis A), cannot be rejected for any player at the 5% level and is rejected for only three players (David Villa, Frank Lampard, and Iker Casillas) at the 10% level.

The table also reports the tests of different subhypotheses concerning whether one's past choices alone, or past opponent's choices alone, or successful past plays alone may determine current choices. No evidence that any player made choices in a serially dependent fashion in any respect is found at the 5% level, and at the 10% level, none of the hypotheses are rejected for more than two players. These results indicate that the choices of most players are unrelated to their own previous choices and outcomes and to their opponents' previous choices and outcomes, exactly as in a random series.

A number of extensions of this investigation are possible. From a more technical perspective, for instance, the statistical power of the tests in various ways, as well as the ability of the tests to detect deviations from minimax play, can be studied using Monte Carlo simulations. From a more empirical perspective, we may consider more strategies such as $C$ and others, and then test the implications in a $3 \times 3$ game or in an $N \times N$ game rather than in a $2 \times 2$ game.[12]

The main finding in this chapter is that the results of the tests are remarkably consistent with equilibrium play in every respect: (1) Winning probabilities are statistically identical across strategies for players, and (2) players generate serially independent sequences and ignore possible strategic links between subsequent penalty kicks. These results, which extend Palacios-Huerta (2003), represent the first time that both implications of von Neumann's (1928) minimax theorem are supported in real life.

*

In recent years, the tests in this chapter have been used to advise a number of teams participating in some of the main club tournaments in the world (e.g., UEFA Champions League and the Football Association Challenge Cup in England, known as the FA Cup), as well as some national teams participating in the top event in the world taking place every four years: the World Cup (Kuper 2011).

In particular, these tests were first used in the UEFA Champions League final on May 21, 2008, in Moscow, to advise Chelsea in its penalty shoot-out versus Manchester United. At the time, no one in the media noticed a pattern in the behavior of the players in the shoot-out, not even a number of small but critical incidents. No one understood what the players were doing and why they were doing it. There was no model to make sense of any behavior. The story is described in great detail in *Soccernomics* (2012) by Simon Kuper and Stefan Szymanski, and this is not the place to repeat it entirely. But it is perhaps worth quoting a few sentences:

> So far, the advice [of the tests] had worked very well for Chelsea (The right-footed penalty-takers had obeyed it to the letter, Manchester United's goalkeeper Van der Sar had not saved a single penalty, and

12 See the appendix in Palacios-Huerta (2003) with some evidence on the $3 \times 3$ game.

> Chelsea's keeper had saved Cristiano Ronaldo's) . . . As Anelka prepared to take Chelsea's seventh penalty, the gangling keeper, standing on the goal-line, extended his arms to either side of him. Then, in what must have been a chilling moment for Anelka, the Dutchman [Van der Sar] pointed with his left hand to the left corner. "That's where you're all putting it, isn't it?" he seemed to be saying. Now Anelka had a terrible dilemma. This was game theory in its rawest form. . . . So Anelka knew that Van der Sar knew that Anelka knew that Van der Sar tended to dive right against right-footers. What was Anelka to do?

You may perhaps know the end. If you do not, this is the authors' summary: "Anelka's decision to ignore the advice [of these tests] probably cost Chelsea the Champions League."

2

# VERNON SMITH MEETS MESSI IN THE LABORATORY

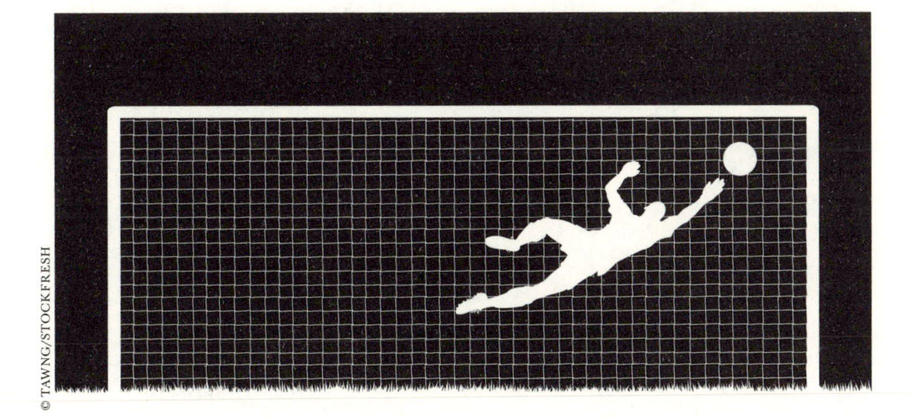



When the exact question being asked and the population being studied are mirrored in a laboratory experiment, the information from the experiment can be clear and informative.

—Armin Falk and James Heckman 2009

A few years ago, a match in the Argentine provinces had to be abandoned just seconds before the finish when the referee, who had just awarded a penalty, was knocked out by an irate player. The league court decided that the last 20 seconds of the game—the penalty kick, in effect—should be played the next Sunday. That gave everyone a week to prepare for the penalty.

At dinner a few nights before the penalty, the goalkeeper, El Gato Díaz, mused about the kicker: "Messi kicks to the right."

"Always," said the president of the club.

"But Messi knows that I know."

"Then, we are fucked."

"Yeah, but I know that he knows," said El Gato.

"Then dive to the left and be ready," said someone at the table.

"No. Messi also knows that I know that he knows," said Gato Díaz, and he got up to go to bed.[1]

We know that some players in real life have gone through the same thought process, which, like El Gato Díaz, they quickly abandoned after realizing they had embarked on an infinite chain of reasoning. After winning the league with Manchester United in the 1993–94 season, Eric Cantona scored two penalties in the 4–0 win in the FA Cup final versus Chelsea. In the press conference after the game, a journalist asked him about his thoughts when he was about to kick his *second* penalty. "Well, I first thought that I would put it in the other corner this time. But I figured the goalkeeper might know this, so I then thought about putting it in the same corner. But then again I thought that he probably thought that I would think like this, so I decided to change again to the other corner. Then I figured that he probably thought that I was thinking that he would probably think that I would think like that, so . . ." After a pause, seemingly realizing that this would take him to an endless chain of reasoning, he concluded: "You know what? The truth is that I just kicked it."

Lionel Messi says that once he has begun his run-up to the ball in a penalty kick, he himself does not know which side he will choose (Sala-i-Martin 2010). The same goes for Franck Ribéry, the designated penalty-taker for Bayern Munich and France. Even after starting on his run-up, he has no idea in which direction he will kick the ball (Kuper and Szymanski 2012). Messi, Ribéry, and others are like coins that do not know on which side they are going to land. As excellent footballers as they all may be in their professional careers, they might do even better as game theoreticians. How about as experimental subjects in a strategic zero-sum game?

The answer to this question goes beyond mere curiosity. It goes to the heart of an important question in economics: How useful are experimental laboratory results? How well do they predict what humans do outside the lab?

Although Vernon Smith received the 2002 Nobel Prize in Economic Sciences "for having established laboratory experiments as a tool in empirical economic analysis," this tool has come under severe attack in recent years. A main critique is that the data generated in laboratory experiments are not realistic, and hence to obtain more realistic data we should pursue experiments not in the lab but in the field.

1 This story is an adaptation of the short story "The Longest Penalty Ever" by Argentine writer Osvaldo Soriano, where Messi replaces Constante, using the translation in Kuper and Szymanski (2012).

Falk and Heckman (2009) explain in some detail why this critique is not only misguided but plain wrong. Consider an outcome of interest $Y$ and a list of determinants $X_1, \ldots, X_N$. Suppose that

$$Y = f(X_1, X_2, \ldots, X_N)$$

Now we are interested in knowing the causal effect of $X_1$ on $Y$, that is, the effect of varying $X_1$, holding fixed $X^* = (X_2, \ldots, X_N)$ . Thus, unless $f$ is *separable* in $X_1$, so that $Y = \theta(X_1) + g(X^*)$, the level of $Y$ response to $X_1$ depends on the level of $X^*$.

Furthermore, even in this separable case, unless $\theta(X_1)$ is a *linear function* of $X_1$, the causal effect of $X_1$ depends on the level of $X_1$ and the size of the variation of $X_1$. These are problems that appear *both* in field and lab experiments, and in any estimation of the causal effect of $X_1$.

$X^*$ may be demographic characteristics, individual preference parameters, social influences, or any set of aspects of the environment. Let $X^*$ denote all these characteristics in a lab setting (say, with student subjects), and $X^{**}$ denote these characteristics in a natural setting (say, with sports card traders as subjects). If one is interested in the causal effect of $X_1$ on $Y$, which one is more informative: holding fixed $X^*$ or holding fixed $X^{**}$?

Experiments are able to obtain universally defined causal effects of $X_1$ on $Y$ only under assumption $Y = \theta(X_1) + g(X^*)$ and only if the response of $Y$ to $X_1$ is linear. But if this is the case, then lab experiments and field experiments are *equally* able to obtain accurate inferences about universal effects. Therefore, the general quest for running experiments in the field to obtain more "realistic" data is fundamentally misguided. In other words, as in the initial quote, if the exact question being asked and the population being studied are exactly mirrored in an experiment, then the information from the experiment can be clear and informative. In fact, it should basically be identical.

Camerer (2011) reviews the available studies on markets, student donations, fishing, grading, sports cards, and restaurant spending that provide the closest matches of lab and field settings, protocols, and subjects and confirms these predictions. He concludes that there is "no replicated evidence that experimental lab data fail to generalize to central empirical features of field data (when the lab features are deliberately closely matched to the field features). . . . The default assumption in the economics profession should be that lab experiments are likely to generalize to closely matched field settings. . . . This is the default assumption, and is generally supported by direct comparisons, in other fields such as biology studies comparing animal behavior in lab settings and in the wild."

The idea in this chapter is to perform the same type of comparative exercise in situations that involve game-theoretical interaction.

This exercise would appear to be particularly relevant in game theory because this is, in fact, one of the main areas where laboratory data are very often used to inform theory.[2] More precisely, this chapter focuses on zero-sum experimental games where players are predicted to choose probability mixtures. Whereas perfectively competitive games do not represent the entire universe of strategic games involving mixed strategies, they are considered a "vital cornerstone" (Aumann 1987) and can be regarded as the branch of game theory with the most solid theoretical foundation.[3]

Soccer has three unique features that make it especially suitable for this purpose:

1. Professional soccer players face a simple strategic interaction that is governed by very detailed rules: a penalty kick;
2. The formal structure of this interaction can be reproduced in the laboratory; and
3. We have seen in the previous chapter that when professional soccer players from European leagues play this game in the field, their behavior is consistent with the *equilibrium* predictions of the theory.

These three distinct characteristics allow us to study whether the skills and heuristics that players may have developed in the field can be transferred to the laboratory, and hence whether laboratory findings are reliable for predicting field behavior in these strategic situations.

The subjects in this chapter were recruited from professional soccer clubs in Spain. As in many other countries, league competition in Spain is hierarchical. It has three professional divisions: Primera División with 20 teams, Segunda División A with 22 teams, and Segunda División B with 80 teams divided into four groups of 20 teams each.[4] The subjects come from clubs in the north of Spain, a region with a high density of professional teams. Eighty male professional soccer players (40 kickers and 40 goalkeepers) are recruited to form 40 kicker–goalkeeper pairs.[5]

2 A reason is that nature rarely creates the circumstances that allow a clear view of the principles at work in strategic situations. See Camerer (2003) for an excellent review.

3 Within the class of zero-sum games, even the less stringent concept of correlated equilibrium coincides with the set of minimax strategies. In this sense, the theory of minimax is one of the less controversial ones from a theoretical point of view.

4 The next division in the hierarchy, Tercera División, has 240 teams and also includes some players who are professional.

5 *Marca* (2012) offers a vitae of every player in the top two divisions. The average age in the sample is 26.5 years, and the average number of years of education is 11.2. No player who had played professionally for less than two years at the time of the experiment was recruited. Data on wages and salaries on individual players are not publicly available, but using Deloitte (2013) information, the average yearly wage may be estimated to be about $1.2 million in the Primera División and about $100,000 in Segunda División A.

Pairs are formed randomly using the last two digits of their national ID cards; the only restriction is that subjects who currently or previously played for the same team were not allowed to participate in the same pair. This measure was implemented to parallel the reality that players encounter in the field. We definitely do not want friends or former teammates to play a game against one another that requires that adversaries get exactly the opposite payoffs.

As indicated in the previous chapter, the clarity and simplicity of the rules in a penalty kick suggest not only that the penalty kick can be studied empirically, but also that it may be easily reproduced in an artificial setting such as a laboratory. The basic structure of a penalty kick may be represented by the simple $2 \times 2$ game already described. This game has a unique Nash equilibrium and requires each player to use a mixed strategy. And as we already know, when the game is repeated, equilibrium theory yields two testable predictions:

1 The probability that a goal will be scored must be the same across each player's strategies and must be equal to the equilibrium scoring probability.
2. Each player's choices must be random. Hence, a player's choices must be independently drawn from a random process and should not depend on his or her own previous play, on the opponent's previous play, or on any other previous actions or outcomes.

The payoffs we use in the experiment are the following:

$$p_{LL} = 0.60, p_{LR} = 0.95, p_{RL} = 0.90, \text{ and } p_{RR} = 0.70$$

which are derived from the empirical probabilities in the sample studied in Palacios-Huerta and Volij (2008). No other field referents or terminology that may trigger any soccer associations were used in the experiment.[6] In particular, subjects were not told that the structure of the game corresponds to a penalty kick or that the payoffs correspond to certain empirically observed probabilities.

The rules of the experiment follow as closely as possible those of O'Neill (1987). The players sit opposite each other at a table. Kickers play the role of row player and goalkeepers the role of column player. Each holds two cards (A and B) with identical backs. A large board across the table prevented them from seeing the backs of each opponent's cards and hence prevented imitation (see the next chapter). The

6 Parameters can add field context to experiments. The idea, pioneered by Hong and Plott (1982) and Grether and Plott (1984), is to estimate parameters that are relevant to field applications and take these into a laboratory setting.

experimenter hands out one page with the following instructions (in Spanish), which he then reads aloud to them:

> We are interested in how people play a simple game. You will first play this game for about 15 hands for practice, just to make sure you are clear about the rules and the results. Then, you will play a series of hands for real money at 1 euro per hand. Before we begin, first examine these dice. They will be used at some point during the experiment. They generate a number between 1 and 100 using a 10-face die for the tens and another 10-face die for the units. The faces of each die are marked from "0" to "9," so the resulting number goes from "00" to "99," where "00" means 100. [The two subjects examine the dice and play with them.] The rules are as follows:
>
> 1. Each player has two cards: A and B.
> 2. When I say "ready" each of you will select a card from your hand and place it face down on the table. When I say "turn," turn your card face up and determine the winner. (I will be recording the cards as played).
> 3. The winner should announce "I win," and will then receive 1 euro.
> 4. Then return the card to your hand, and get it ready for the next round.
>
> I will explain how the winner is determined next. Are there any questions so far?
>
> Now, the winner is determined with the help of the dice as follows:
>
> - If there is a match *AA*, [row player's name] wins if the dice yield a number between 01 and 60; otherwise [column player's name] wins.
> - If there is a match *BB*, [row player's name] wins if the dice yield a number between 01 and 70; otherwise [column player's name] wins.
> - If there is a mismatch *AB*, [row player's name] wins if the dice yield a number between 01 and 95; otherwise [column player's name] wins.
> - If there is a mismatch *BA*, [row player's name] wins if the dice yield a number between 01 and 90; otherwise [column player's name] wins.
>
> The following diagram may be useful:

| | *A* | *B* |
|---|---|---|
| *A* | 0.60 | 0.95 |
| *B* | 0.90 | 0.70 |

> Are there any questions?

Thus, the game was presented with the help of a matrix, and subjects learned the rules through a few rounds of practice. The unique

**A. Frequencies**

| | | Column Player Choice: L | Column Player Choice: R | Marginal Frequecies for Row Player |
|---|---|---|---|---|
| Row Player Choice | L | 0.152<br>(0.165)<br>[0.0068] | 0.182<br>(0.198)<br>[0.0073] | 0.333<br>(0.364)<br>[0.0088] |
| | R | 0.310<br>(0.289)<br>[0.0083] | 0.356<br>(0.347)<br>[0.0087] | 0.667<br>(0.636)<br>[0.0088] |
| Marginal Frequencies for Column Player | | 0.462<br>(0.455)<br>[0.009] | 0.538<br>(0.545)<br>[0.009] | |

**B. Win Percentages**

| | |
|---|---|
| **Observed Row Player Win Percentage:** | 0.7947 |
| **Minimax Row Player Win Percentage:** | 0.7909 |
| **Minimax Row Player Win Std. Deviation:** | 0.0074 |

**Figure 2.1.** Relative frequencies of choices and win percentages in penalty kick experiment for professional players.

mixed-strategy equilibrium of this game dictates that row and column players choose the *A* card with probabilities 0.3636 and 0.4545, respectively. The subjects played 15 rounds for practice and then 150 times for real money, proceeding at their own pace. They were not told the number of hands they would play. On the few occasions when they made an error announcing the winner, the experimenter corrected them. A typical session lasted about 70 minutes.

Figure 2.1 presents aggregate statistics describing the outcomes of the experiment using the standard notation of *L* and *R* instead of *A* and *B*. In the top panel, each interior cell reports the relative frequency with which the pair of moves corresponding to that cell occurred. The minimax relative frequencies appear in parentheses, and the standard deviation for the observed relative frequencies under the minimax hypothesis appear in brackets. At the bottom and to the right are the overall relative frequencies with which players were observed to play a particular card, again accompanied by the corresponding relative frequencies and standard deviations under the minimax model. Observed and minimax win frequencies for the row player are reported in the bottom panel.

The pattern of observed relative frequencies for each pair of choices shows a general consistency with the minimax model in that they all are within 1 to 2 percentage points from the predicted frequencies. Likewise, the marginal frequencies of actions for the column players are extremely close (in fact, statistically identical) to the minimax predictions. Row players, on the other hand, choose frequencies 0.333 for $L$ and 0.667 for $R$, which, though close to the minimax predictions of 0.363 and 0.637, are statistically different from them.[7] The observed aggregate row player win frequency (0.7947) is less than one standard deviation away from the theoretically expected value (0.7909). Hence, although the aggregate mixture of the row players is statistically different from the equilibrium one, the difference is minuscule. Indeed, row players chose $L$ with probability 0.33, and the equilibrium prescribes 0.36. Also, if column players played the best response to row players' actual mixes, their success rates would increase from 20.9% to only 21.6%, an arguably ridiculously tiny amount.

Data at the individual pair level allow a closer scrutiny of the extent to which minimax play may be supported for most individual subjects and most pairs of players. Table 2.1 reports the relative frequencies of choices for each of the 20 pairs in the sample as well as some initial tests of the model.

The minimax hypothesis implies that the choices of actions represent *independent* drawings from a binomial distribution where the probabilities of $L$ are 0.363 and 0.454 for the row and column players, respectively. We should then expect a binomial test of conformity with minimax play to reject the null hypothesis for 2 and 4 players at the 5% and 10% significance levels. The results show that in fact these are precisely the number of rejections at those confidence levels.

Thus, these initial findings support the hypothesis that professional soccer players play very close to the equilibrium of the game, though not perfectly, in a laboratory. However, since equilibrium behavior also implies that action combinations should be realizations of *independent* drawings of a multinomial distribution, further support is needed. To test whether the players' actions are correlated, we perform the following test.

7 The $p$-value of the null hypothesis that players choose the equilibrium frequencies is 0.06% for row players, 0.41% for column players, and 0.48% for both players.

**Table 2.1.** Marginal Frequencies and Action Pair Frequencies in Penalty Kick's Experiment for Professional Players

| | Marginal Frequencies | | Pair Frequencies | | | | |
|---|---|---|---|---|---|---|---|
| Pair # | Row *L* | Column *L* | *LL* | *LR* | *RL* | *RR* | $\chi^2$ *p*-value |
| 1 | 0.320 | 0.453 | 0.140 | 0.180 | 0.313 | 0.367 | 0.729 |
| 2 | 0.360 | 0.380* | 0.127 | 0.233 | 0.253 | 0.387 | 0.305 |
| 3 | 0.307 | 0.427 | 0.127 | 0.180 | 0.300 | 0.393 | 0.459 |
| 4 | 0.327 | 0.460 | 0.153 | 0.173 | 0.307 | 0.367 | 0.819 |
| 5 | 0.327 | 0.493 | 0.153 | 0.173 | 0.340 | 0.333 | 0.568 |
| 6 | 0.340 | 0.480 | 0.140 | 0.200 | 0.340 | 0.320 | 0.525 |
| 7 | 0.287** | 0.427 | 0.133 | 0.153 | 0.293 | 0.420 | 0.190 |
| 8 | 0.320 | 0.460 | 0.100 | 0.220 | 0.360 | 0.320 | 0.068* |
| 9 | 0.307 | 0.467 | 0.133 | 0.173 | 0.333 | 0.360 | 0.479 |
| 10 | 0.313 | 0.480 | 0.167 | 0.147 | 0.313 | 0.373 | 0.454 |
| 11 | 0.353 | 0.480 | 0.180 | 0.173 | 0.300 | 0.347 | 0.866 |
| 12 | 0.427* | 0.480 | 0.193 | 0.233 | 0.287 | 0.287 | 0.359 |
| 13 | 0.367 | 0.473 | 0.167 | 0.200 | 0.307 | 0.327 | 0.952 |
| 14 | 0.327 | 0.447 | 0.153 | 0.173 | 0.293 | 0.380 | 0.782 |
| 15 | 0.340 | 0.553** | 0.173 | 0.167 | 0.380 | 0.280 | 0.071* |
| 16 | 0.320 | 0.473 | 0.160 | 0.160 | 0.313 | 0.367 | 0.659 |
| 17 | 0.347 | 0.467 | 0.200 | 0.147 | 0.267 | 0.387 | 0.256 |
| 18 | 0.327 | 0.440 | 0.140 | 0.187 | 0.300 | 0.373 | 0.791 |
| 19 | 0.327 | 0.440 | 0.140 | 0.187 | 0.300 | 0.373 | 0.791 |
| 20 | 0.327 | 0.460 | 0.153 | 0.173 | 0.307 | 0.367 | 0.819 |

*Notes*: ** and * denote rejections at the 5% and 10% levels, respectively, of the minimax model for individual players. In the last column, they denote rejection for the joint hypothesis that both players choose actions with the equilibrium frequencies.

Minimax play implies that action combinations are realizations of independent drawings from a multinomial distribution with probabilities 0.165 for *LL*, 0.198 for *LR*, 0.289 for *RL*, and 0.347 for *RR*. The table reports the relative frequencies of each combination of actions for each of the pairs in the sample. Using the corresponding absolute frequencies along with their minimax probabilities, we can then test the joint hypothesis that players choose actions with the equilibrium frequency *and* that their choices are stochastically independent. This is a $\chi^2$ test with three degrees of freedom. Using the 5% and 10% levels of significance, when play follows minimax, we would expect to reject the null hypothesis for 1 and 2 pairs, respectively. We find 0 and 2 rejections, respectively.

Summing up, even though the observed aggregate frequency for the row players is statistically different from the equilibrium predictions, the evidence thus far lends substantial support to the minimax hypothesis. Our next task is to test more closely the implications of the equilibrium of the game.

### IMPLICATION 1: WINNING RATES AND THE DISTRIBUTION OF PLAY

Minimax play implies that the success probabilities of each action are the same for each player and are equal to 0.7909 for the row player and 0.2090 for the column player. Furthermore, when combined with the equilibrium strategies, we can obtain the relative action–outcome frequencies associated with the equilibrium. Table 2.2 reports the relative frequencies of action–outcome combinations observed for each of the row and column players in the sample. Using the absolute frequencies corresponding to these entries, we can then implement a $\chi^2$ test of conformity with minimax play. This test would be identical to the one performed previously if it were not for the fact that the success rate is determined not only by the choice of strategies but also by the realization of the dice.

The results of the test show that the null hypothesis is rejected for no player at the 5% significance level, and for three players at the 10% level; both cases are fewer than the expected number of rejections, 2 and 4, respectively. Hence, at the individual level, the hypothesis that scoring probabilities are identical both across strategies and to the equilibrium rate cannot be rejected for most players at conventional significance levels.

If we move from the individual to the aggregate level, we can test the joint hypothesis that each one of the experiments is simultaneously generated by equilibrium play. The test statistic is simply the sum of the individual test statistics. Under the null hypothesis, it is distributed as a $\chi^2$ with 60 degrees of freedom both for the set of row players and for the set of column players. We find that the Pearson statistics are 40.002 and 32.486, with an associated $p$-value above 90% in both cases.[8] Hence, the null hypothesis that the data for all players are generated by equilibrium play cannot be rejected at conventional significance levels.

Therefore, these individual and aggregate results are quite consistent with the hypothesis that these professional players equate their strategies' payoffs to the equilibrium ones in a laboratory.

8 The test statistics for the row and column players may not be added, given that within each pair the players' success rates are not independent.

## IMPLICATION 2: THE RANDOMNESS HYPOTHESIS

The second testable implication of equilibrium play is that a player should randomize using the same distribution at each stage of the game. This notion implies that players' choices are serially independent. This hypothesis has never found support in a laboratory setting, although to the best of our knowledge it has always been studied with inexperienced players.[9] Therefore, the novel question we can address is whether the skills and experience of professional players randomizing in the field are useful to generate random sequences in a lab setting. To study this question, we report two tests. First, the "runs test," which, as indicated earlier, is a test with low power to detect deviations from randomization in that subjects may behave nonrandomly and this test may not be able to detect it. Second, we also perform a logit regression test based on the Arellano–Carrasco model with the same specification as in the previous chapter.

It turns out that the null hypothesis of serial independence is rejected in both tests for two players at the 5% significance level and for four players at the 10% level. This result corresponds precisely to the expected number of rejections in both cases under the assumption that the null hypothesis is true. In other words, the hypothesis that professional soccer players generate random sequences in a laboratory cannot be rejected: They neither switch strategies too often nor too little, the number of rejections is remarkably consistent with the theory, and there is no lagged variable in the AC regression that helps predict current choices.

Overall, the behavior of professional soccer players contrasts sharply with the overwhelming experimental evidence from the psychological and economics literatures. Years of field experience appears to be quite valuable for generating randomness in strategically similar laboratory games.[10] This is in fact the first time that subjects have been found to display statistically significant serial independence in a strategic laboratory game. Together with the result that subjects equate payoffs across strategies and to the equilibrium success rates, the behavior of these professional players represents the first time that a pool of subjects satisfies the equilibrium conditions in the laboratory in games requiring probabilistic mixtures.

9 As noted in the first chapter, the only exception is Neuringer (1986), in which subjects were explicitly taught to choose randomly after hours of training by providing them with detailed feedback from previous responses in the experiment.

10 The experience not only comes from penalty kicks (repeated infrequently) but also from situations where repetitions may be taken in rapid succession (e.g., dribbling, passes, or corners) that also require unpredictability.

**Table 2.2.** Tests of Equality of Payoffs and Randomization Tests

| | | $L$ | | $R$ | | Pearson Tests | | Runs Tests | | | |
|---|---|---|---|---|---|---|---|---|---|---|---|
| Pair # | Player | Success | Fail | Success | Fail | Statistic | $p$-value | $r$ | $\Phi[r-1,s]$ | $\Phi[r,s]$ | Arellano–Carrasco Tests |
| 1 | Row | 0.260 | 0.060 | 0.540 | 0.140 | 1.360 | 0.715 | 72 | 0.840 | 0.877 | |
| | Column | 0.080 | 0.373 | 0.120 | 0.427 | 0.491 | 0.921 | 69 | 0.129 | 0.167 | |
| 2 | Row | 0.300 | 0.060 | 0.500 | 0.140 | 0.645 | 0.886 | 74 | 0.727 | 0.779 | |
| | Column | 0.047 | 0.333 | 0.153 | 0.467 | 6.441 | 0.092* | 72 | 0.488 | 0.554 | |
| 3 | Row | 0.233 | 0.073 | 0.553 | 0.140 | 2.351 | 0.503 | 64 | 0.404 | 0.469 | |
| | Column | 0.100 | 0.327 | 0.113 | 0.460 | 0.774 | 0.856 | 82 | 0.884 | 0.913 | |
| 4 | Row | 0.247 | 0.080 | 0.540 | 0.133 | 1.306 | 0.728 | 69 | 0.604 | 0.682 | |
| | Column | 0.107 | 0.353 | 0.107 | 0.433 | 0.302 | 0.960 | 75 | 0.433 | 0.499 | |
| 5 | Row | 0.280 | 0.047 | 0.520 | 0.153 | 2.278 | 0.517 | 79 | 0.985** | 0.992 | AC |
| | Column | 0.100 | 0.393 | 0.100 | 0.407 | 0.989 | 0.804 | 80 | 0.717 | 0.770 | |
| 6 | Row | 0.280 | 0.060 | 0.513 | 0.147 | 0.776 | 0.855 | 74 | 0.830 | 0.869 | |
| | Column | 0.080 | 0.400 | 0.127 | 0.393 | 1.755 | 0.625 | 89 | 0.981** | 0.987 | AC |
| 7 | Row | 0.207 | 0.080 | 0.600 | 0.113 | 6.673 | 0.083* | 53 | 0.025 | 0.041* | |
| | Column | 0.093 | 0.333 | 0.100 | 0.473 | 1.161 | 0.762 | 72 | 0.315 | 0.375 | |
| 8 | Row | 0.273 | 0.047 | 0.507 | 0.173 | 3.640 | 0.303 | 69 | 0.655 | 0.730 | |
| | Column | 0.113 | 0.347 | 0.107 | 0.433 | 0.670 | 0.880 | 69 | 0.124 | 0.160 | |
| 9 | Row | 0.233 | 0.073 | 0.560 | 0.133 | 2.508 | 0.474 | 63 | 0.323 | 0.404 | |
| | Column | 0.113 | 0.353 | 0.093 | 0.440 | 1.134 | 0.769 | 67 | 0.066 | 0.089 | |

| | | | | | | | | | | | |
|---|---|---|---|---|---|---|---|---|---|---|---|
| 10 | Row | 0.247 | 0.067 | 0.560 | 0.127 | 2.051 | 0.562 | 58 | 0.065 | 0.089 | |
| | Column | 0.093 | 0.387 | 0.100 | 0.420 | 0.617 | 0.892 | 85 | 0.922 | 0.943 | |
| 11 | Row | 0.260 | 0.093 | 0.513 | 0.133 | 1.018 | 0.797 | 66 | 0.235 | 0.289 | |
| | Column | 0.107 | 0.373 | 0.120 | 0.400 | 0.683 | 0.877 | 69 | 0.113 | 0.147 | |
| 12 | Row | 0.327 | 0.100 | 0.493 | 0.080 | 5.132 | 0.162 | 68 | 0.125 | 0.162 | |
| | Column | 0.073 | 0.407 | 0.107 | 0.413 | 1.857 | 0.603 | 77 | 0.541 | 0.605 | |
| 13 | Row | 0.287 | 0.080 | 0.480 | 0.153 | 0.657 | 0.883 | 71 | 0.484 | 0.559 | |
| | Column | 0.100 | 0.373 | 0.133 | 0.393 | 1.112 | 0.774 | 80 | 0.729 | 0.781 | |
| 14 | Row | 0.247 | 0.080 | 0.553 | 0.120 | 1.843 | 0.606 | 72 | 0.802 | 0.845 | AC |
| | Column | 0.080 | 0.367 | 0.120 | 0.433 | 0.426 | 0.935 | 63 | 0.018 | 0.027* | |
| 15 | Row | 0.260 | 0.080 | 0.533 | 0.127 | 0.743 | 0.863 | 68 | 0.441 | 0.507 | |
| | Column | 0.093 | 0.460 | 0.113 | 0.333 | 7.563 | 0.056* | 68 | 0.103 | 0.135 | |
| 16 | Row | 0.253 | 0.067 | 0.553 | 0.127 | 1.578 | 0.664 | 67 | 0.509 | 0.592 | |
| | Column | 0.073 | 0.400 | 0.120 | 0.407 | 1.687 | 0.640 | 74 | 0.353 | 0.416 | |
| 17 | Row | 0.253 | 0.093 | 0.540 | 0.113 | 2.043 | 0.564 | 71 | 0.605 | 0.679 | |
| | Column | 0.120 | 0.347 | 0.087 | 0.447 | 2.119 | 0.548 | 72 | 0.246 | 0.301 | |
| 18 | Row | 0.253 | 0.073 | 0.533 | 0.140 | 0.950 | 0.813 | 62 | 0.156 | 0.199 | |
| | Column | 0.087 | 0.353 | 0.127 | 0.433 | 0.337 | 0.953 | 71 | 0.231 | 0.285 | |
| 19 | Row | 0.260 | 0.067 | 0.527 | 0.147 | 0.942 | 0.815 | 68 | 0.539 | 0.604 | |
| | Column | 0.073 | 0.367 | 0.140 | 0.420 | 1.696 | 0.638 | 78 | 0.666 | 0.724 | |
| 20 | Row | 0.260 | 0.067 | 0.553 | 0.120 | 1.509 | 0.680 | 75 | 0.918 | 0.947 | AC |
| | Column | 0.093 | 0.367 | 0.093 | 0.447 | 0.671 | 0.880 | 71 | 0.204 | 0.254 | |

*Notes*: ** and * denote rejections at the 5% and 10% levels, respectively. AC denotes the cases in which subjects did not pass the test null hypothesis A in table 1.3.

*

Two hundred and seventy five years ago, David Hume (1739) stressed that

> We transfer our experience in past instances to objects which are resembling, but are not exactly the same with those concerning which we have had experience. . . . Tho' the habit loses somewhat of its force by every difference, yet 'tis seldom entirely destroy'd where any considerable circumstances remain the same.

In zero-sum games, when "considerable circumstances remain the same," professional soccer players confirm Hume's insight. Laboratory experiments not only permit valuable control of players' information, payoffs, available strategies, and other relevant aspects, but they also generate the same information as real life in these games. The next chapter studies some situations in the same setting and for the same type of players where important circumstances do not remain the same. We would then expect behavior not to remain the same.

A final point concerns other species. In the animal kingdom, randomness can be advantageous to an animal if it is unpredictable enough to confuse its enemies (Driver and Humphries 1988; Miller 1997). Research in evolution has documented that adaptively unpredictable behavior, or "Protean behavior," is often observed in wildlife. The term "Protean" comes from the mythical Greek river-god Proteus, who eluded capture by continually and unpredictably changing form. Some animals have even been taken to the lab to study their behavior in competitive games involving mixed strategies. Remarkably, Martin et al. (2013) show that experienced chimpanzees appear to behave very close to equilibrium behavior in these games in the laboratory, in fact, much closer than standard inexperienced humans. Their interpretation is that "the results are generally consistent with the cognitive tradeoff hypothesis, which conjectures that some human cognitive ability inherited from chimpanzee kin may have been displaced by dramatic growth in the human neural capacity for language. As a result, chimpanzees retained the ability, slightly superior to humans, to adjust strategy competitively and in unpredictable ways." Indeed, competitive ability and short-term memory appear to be "preserved" in chimpanzee evolution (and also practiced a lot when the chimps are young) but displaced in the human cortex by language and coordination-specific skills.

When we look at experienced humans and match data from actual play on the field with laboratory data of players from the same group (professionals in European leagues), the results show that the human species has no reason to envy chimpanzees, at least in this competitive dimension.

# 3

# LESSONS FOR EXPERIMENTAL DESIGN

Nemo me impune lacessit.
—Motto of the Order of the Thistle, Scotland

The playing of games is dependent on abilities that game theory does not capture well, such as memory, the ability to process information and the quality of associations. The assimilation of these concepts constitutes one of the main challenges for the future.
—Ariel Rubinstein 2004

We learned in chapter 2 that when the exact question being asked is mirrored in a laboratory experiment and the population being studied is the same as in the field, the outcomes from the experiment can be just as clear and informative. This result suggests that when either the exact question being asked is not mirrored or the population being studied differs, the outcomes from the experiment probably do not parallel those observed in the field. Here we use this insight to draw four lessons for experimental design using the games, methods, and results from the previous chapters.

*

## LESSON 1

We have seen how players from the top professional soccer leagues in the world (Spain, United Kingdom, Italy, and others) play minimax in the field. This evidence, however, does not necessarily mean that all soccer professionals everywhere in the world also play minimax. There are powerful reasons to think that this is probably not the case.

First, European professional leagues are more than 100 years old and extremely competitive. Each league has, across all the different hierarchical divisions, more than 100 professional teams, which typically have their own academies with several amateur teams. Most other leagues in the world do not come close to their level of competitiveness and professionalism.

Second, skill acquisition requires considerable investment, and hence players in leagues that are younger, less established, or less competitive than the European leagues may not have had sufficient time and incentives to develop the relevant skill portfolio.

Third, strategic skills are special in the sense that one's own optimal behavior depends on the opponent's behavior too. For instance, for minimax skills to develop in the different zero-sum situations that players face in the field, it is *necessary* that *all* subjects play (or expect others to play) according to equilibrium behavior. If players do not believe that their opponents are playing minimax, that is, if their opponents systematically deviate from minimax or are expected to do so, then minimax play is no longer optimal. This in turn means that they probably do not develop the appropriate minimax skills.

These types of considerations suggest that it would be interesting to look at the testable implications of minimax in a professional league that is at the other extreme of the distribution in terms of age, experience, and competitiveness: a young league, not as competitive, formed by mostly relatively inexperienced players.

Consider Major League Soccer (MLS) in the United States. Very few players from the MLS league, even MLS superstars, have ever attained a level that would qualify them for play in second division leagues in Europe. Take, for instance, some of the best American players ever, MLS All-Stars and MLS All-Time Best, such as Brian McBride, Claudio Reyna, and Freddy Adu. The first two struggled to play with first and second division teams in Germany, whereas Adu did not see much action in his team in the Portuguese league and was loaned to lower teams several times. He was recently seen on loan in the Turkish second division side Çaykur Rizespor. The MLS is a young league that was created in 1996 with just 10 teams, and now it has 19 teams. Most of the teams are composed of a large proportion of inexperienced players coming out of college. As such, the MLS is considered to be "low quality" even by their own governance. Probably as a result of this low quality, most clubs systematically lose money, which in turn means that the league is in serious financial difficulties. With the exception of one or two teams in recent years, every team has lost money every year since the league's creation. Although the teams are slowly improving, the league is still considered to be low level by international standards. Over the past few years, Commissioner Don Garber has publicly voiced concerns about the league's quality and implemented a number of changes to focus on improving the quality of play its teams produce in the field, including the Designated Player rule and the creation of a leaguewide youth development system. These characteristics make the MLS a good candidate to study players from a league at the other end of the distribution of quality.

Table 3.1 reports evidence of the same tests as in previous chapters with real-life data for 20 players from the MLS. The evidence shows that exactly *half* of the sample does *not* equate payoffs across strategies (minimax implication number 1) at the 10% level when we would only expect two rejections in equilibrium. At the aggregate level, the $\chi^2$ test has a $p$-value of $1.03 \times 10^{-3}$, indicating that the hypothesis that *all of them* equate payoffs across strategies can be rejected at virtually any confidence level. The results of the randomization tests (both the runs tests and the Arellano–Carrasco tests) for minimax implication number 2 go in the same direction: at the 10% level, 12 and 14 of the 20 players fail to pass the runs test and the Arellano–Carrasco test, when we would again expect just two rejections in equilibrium.

The results of an interesting survey conducted on MLS players confirm the non-minimax behavior we just observed in the field. When MLS players are asked how they like to kick a penalty kick, a remarkable 44% of the subjects report playing, or that they would play, *pure* strategies in a penalty kick (see table 3.2). That is, quite shockingly, they declare that they would *always* kick to the *same* place!

Thus, MLS players do not play minimax in the field, and they say that they do not and would not play minimax in the field. In contrast,

**Table 3.1.** Tests of Equality of Payoffs and Randomization for MLS Players

| | | Choices | | Win Rates | | Pearson Tests | | Runs Tests | | | |
|---|---|---|---|---|---|---|---|---|---|---|---|
| Player | $N$ | $L$ | $R$ | $L$ | $R$ | Statistic | $p$-value | $r$ | $\Phi[r-1,s]$ | $\Phi[r,s]$ | Arellano–Carrasco Tests |
| 1 | 31 | 5 | 26 | 0.800 | 0.384 | 2.921 | 0.087* | 12 | 0.929 | 0.984 | |
| 2 | 28 | 7 | 21 | 0.571 | 0.523 | 0.047 | 0.826 | 13 | 0.698 | 0.850 | AC |
| 3 | 26 | 3 | 23 | 1.000 | 0.695 | 1.249 | 0.263 | 9 | 0.989** | 0.999 | AC |
| 4 | 32 | 6 | 26 | 0.833 | 0.384 | 3.941 | 0.047** | 9 | 0.087 | 0.225 | |
| 5 | 32 | 7 | 25 | 0.857 | 0.600 | 1.602 | 0.205 | 16 | 0.971* | 0.992 | AC |
| 6 | 37 | 12 | 25 | 0.833 | 0.880 | 0.151 | 0.697 | 11 | 0.005 | 0.014** | AC |
| 7 | 40 | 13 | 27 | 0.692 | 0.666 | 0.026 | 0.871 | 16 | 0.131 | 0.226 | AC |
| 8 | 34 | 7 | 27 | 0.857 | 0.407 | 4.497 | 0.033** | 17 | 0.991** | 0.998 | AC |
| 9 | 38 | 14 | 24 | 0.428 | 0.750 | 3.926 | 0.047** | 12 | 0.005 | 0.014** | |
| 10 | 50 | 9 | 41 | 0.888 | 0.707 | 1.264 | 0.260 | 10 | 0.001 | 0.004** | AC |
| 11 | 43 | 11 | 32 | 0.818 | 0.750 | 0.213 | 0.644 | 24 | 0.993** | 0.998 | AC |
| 12 | 30 | 6 | 24 | 0.833 | 0.583 | 1.291 | 0.255 | 17 | 0.999** | 0.999 | AC |
| 13 | 35 | 7 | 28 | 0.857 | 0.642 | 1.193 | 0.274 | 17 | 0.990** | 0.998 | AC |
| 14 | 34 | 8 | 26 | 0.875 | 0.538 | 2.933 | 0.086* | 18 | 0.981** | 0.995 | AC |
| 15 | 30 | 6 | 24 | 0.500 | 0.666 | 0.574 | 0.448 | 8 | 0.033 | 0.106 | AC |
| 16 | 28 | 14 | 14 | 0.357 | 0.785 | 5.250 | 0.021** | 21 | 0.982** | 0.993 | |
| 17 | 32 | 6 | 26 | 0.666 | 0.923 | 2.930 | 0.086* | 8 | 0.025 | 0.087 | |
| 18 | 33 | 9 | 24 | 0.888 | 0.583 | 2.750 | 0.097* | 12 | 0.122 | 0.237 | AC |
| 19 | 40 | 8 | 32 | 0.875 | 0.718 | 4.833 | 0.027** | 9 | 0.003 | 0.014** | AC |
| 20 | 28 | 9 | 19 | 0.666 | 0.789 | 3.602 | 0.057* | 11 | 0.114 | 0.223 | |
| All | 681 | | | | | 45.201 | 0.001*** | | | | |

*Notes*: ***, **, and * denote rejections at the 1%, 5%, and 10% levels, respectively. AC denotes rejection of randomization using the Arellano–Carrasco (2003) test.

**Table 3.2.** Proportion of Professional Players Who Indicate That They Would Always Kick to the Same Side in a Real-Life Penalty Kick

| | United States* | England | France | Spain | Germany |
|---|---|---|---|---|---|
| Professional Players | 44% ($N = 20$) | 0% ($N = 24$) | 0% ($N = 30$) | 0% ($N = 42$) | 0% ($N = 57$) |
| Amateur Players | – | 0% ($N = 22$) | 2% ($N = 48$) | 2% ($N = 50$) | 0% ($N = 51$) |

*US data come from the survey reported in the working paper version of Levitt et al. (2010). The data for England, France, Spain, and Germany come from the same survey implemented in these countries. No professional player in the survey had participated before in a laboratory experiment.

when the same survey is conducted on players from the top European leagues, not a single professional answers that he would always kick to the same place.

Minimax strategic skills are special and, like other human skills, are rarely developed unless the setting calls for them. From the perspective of experimental studies of zero-sum games, the lesson is that MLS players would not be an appropriate pool of subjects to conduct the type of study implemented in the previous chapter.[1] Laboratories are not magical settings, and there is no reason why behavior that is not observed in the field should suddenly materialize in the lab.

## LESSON 2

Some experimenters consider that "in an attempt to ensure generality and control by gutting all instructions and procedures of field referents, the traditional lab experimenter has arguably lost control to the extent that subjects seek to provide their own field referents. The obvious solution is to conduct experiments both ways: with and without naturally occurring field referents and context" (Harrison and List 2004, p. 1050).

We take this insight seriously in this and the next lesson by deliberately "losing control." The idea is to lose two naturally occurring field referents or contexts that may have helped subjects perceive, consciously or unconsciously, the laboratory game in the previous chapter as being

1 Neither would players in the Israeli league. Bar-Eli et al. (2007) find that although in the Israeli league minimax fits the data better than other alternative theories, players are still far from the minimax predictions. The Israeli league is, like the MLS, a low-quality league with mostly inexperienced players.

as strictly competitive as the situations they face in their professional lives. We then see whether we find substantial differences in behavior.

We follow exactly the same experimental design as in chapter 2, including payoffs, instructions, repetitions, and all other characteristics. Players are also recruited from the professional soccer leagues in Spain, and the new subjects have neither participated in nor heard about the experiments in chapter 2.[2]

The first "artificial" margin that we consider departs from the original design in chapter 2 and from the conditions that subjects face in real life in chapter 1 in that we allow "friendship" to play a role. We ask *friends* (two players from the same team), rather than *enemies* (two players from different teams who are not friends and have never met before) to play against one another in the same pair.[3]

Soccer is a team sport, and the job of a player is to cooperate with his or her teammates for the common good of the team. Teammates are used to cooperating, which suggests that a laboratory zero-sum game between friends may well be embedded in a "bigger game," one that does not exactly correspond to a zero-sum game. Such considerations could shatter minimax play.

We recruit 30 players to form 15 kicker–goalkeeper pairs, as in the previous chapter, but the kicker and goalkeeper are from the *same* team. We then ask these pairs of *teammates* to play the same penalty kick game.

With 30 subjects, the expected number of rejections in equilibrium is just 1.5 and 3 at the 5% and 10% significance levels (see table 3.3). We find that the actual number of players for whom the null hypothesis is rejected is 5 and 10 at these levels for the Pearson test. That is, the number of rejections is more than three times greater than the expected number in equilibrium, and hence an order of magnitude greater than the proportions of rejections obtained in chapter 2 when the game is played by subjects who are not and had never been friends. In addition, at the aggregate level, the Pearson test conclusively rejects ($p$-value $< 0.01$) the hypothesis that all players equate payoffs across strategies. Finally, even greater discrepancies relative to the equilibrium are found in the AC randomization tests.

We conclude from this experiment that *friends* do not behave like *enemies*. Yes, I know this conclusion probably comes as a terrible shock, but dreary data show that it is true. Still it makes an important point:

2 The experiments were conducted in the Universidad del País Vasco and at the academy of Athletic Club de Bilbao, Spain.

3 In professional leagues, players from different teams of course typically know each other, even though they are not friends.

The game that friends play appears to be embedded in a larger game not captured by the experimental design. The game (a zero-sum situation played among friends) does not represent the way subjects interact in the field. Since friends cooperate, the very control of having friends play against one another in a lab experiment is an artificial margin that causes deviations from equilibrium in a game that requires strict, fierce competition.

### LESSON 3

In real-life penalty kicks, the goalkeeper's ultimate adversary is the kicker, just as the kicker would like to wish the goalkeeper away. Having a kicker in front of a goalkeeper and vice versa in an experimental study of a strictly competitive game, as in chapter 2, may readily provide relevant context (e.g., a penalty kick) and trigger very different responses than the same game played between players of the same type. We next study the role of this field referent.

We recruit a different set of 30 players to form 15 *same-type* pairs, that is, kicker–kicker or goalkeeper–goalkeeper, with each paired subject from a different team. As table 3.4 shows, we find that the number of players for whom the null hypothesis is rejected is 5 and 8 at the 5% and 10% significance levels. Therefore, the number of rejections is much greater than the expected number in equilibrium. And again, at the aggregate level the Pearson test conclusively rejects ($p$-value $< 0.01$) the hypothesis that all players equate payoffs across strategies, and similarly large discrepancies are found in the randomization tests.

Thus, having a kicker play versus a goalkeeper a strictly competitive game triggers, both in real life and in the lab, very different reactions than having two players of the same type face each other in a lab. Matching a kicker with a goalkeeper in the lab appears to provide subjects a naturally occurring referent that contributes to determining whether minimax behavior obtains.

Lessons 2 and 3 accord well with intuition, which suggests that laboratory studies of strictly competitive games (and possibly all other games) should benefit from capturing the fundamental competitive conditions that subjects encounter in real life. Together with Lesson 1, the results suggest that what does not happen in the field should not and does not happen in the lab.[4]

4 These three lessons confirm the findings in Levitt et al. (2010).

**Table 3.3.** Tests of Equality of Payoffs and Randomization Tests with Friends

| | | Frequencies | | Win Rates | | Pearson Tests | | Runs Tests | | | |
|---|---|---|---|---|---|---|---|---|---|---|---|
| Pair # | Player | $L$ | $R$ | $L$ | $R$ | Statistic | $p$-value | $r$ | $\Phi[r-1,s]$ | $\Phi[r,s]$ | Arellano–Carrasco Tests |
| 1 | K | 0.589 | 0.411 | 0.659 | 0.790 | 3.063 | 0.080* | 72 | 0.352 | 0.416 | |
| | G | 0.612 | 0.388 | 0.185 | 0.448 | 12.077 | 0.000** | 71 | 0.388 | 0.455 | |
| 2 | K | 0.591 | 0.409 | 0.809 | 0.738 | 1.071 | 0.300 | 58 | 0.005 | 0.009** | AC |
| | G | 0.607 | 0.393 | 0.176 | 0.288 | 2.631 | 0.104 | 66 | 0.111 | 0.147 | |
| 3 | K | 0.587 | 0.413 | 0.693 | 0.597 | 1.492 | 0.221 | 94 | 0.999** | 0.999 | AC |
| | G | 0.613 | 0.387 | 0.304 | 0.414 | 1.881 | 0.170 | 58 | 0.005 | 0.009** | AC |
| 4 | K | 0.600 | 0.400 | 0.856 | 1.000 | 9.489 | 0.002** | 79 | 0.826 | 0.866 | AC |
| | G | 0.599 | 0.401 | 0.089 | 0.083 | 0.014 | 0.905 | 80 | 0.866 | 0.899 | AC |
| 5 | K | 0.395 | 0.605 | 0.915 | 0.923 | 0.029 | 0.863 | 82 | 0.937 | 0.955 | AC |
| | G | 0.694 | 0.306 | 0.067 | 0.109 | 0.742 | 0.388 | 59 | 0.112 | 0.153 | |
| 6 | K | 0.449 | 0.551 | 0.940 | 0.976 | 1.223 | 0.268 | 76 | 0.523 | 0.588 | |
| | G | 0.607 | 0.393 | 0.033 | 0.051 | 0.298 | 0.585 | 69 | 0.241 | 0.298 | AC |
| 7 | K | 0.417 | 0.583 | 0.651 | 0.667 | 0.041 | 0.839 | 63 | 0.025 | 0.037* | AC |
| | G | 0.603 | 0.397 | 0.289 | 0.417 | 2.619 | 0.105 | 83 | 0.947 | 0.963 | AC |
| 8 | K | 0.580 | 0.420 | 0.713 | 0.683 | 0.157 | 0.691 | 68 | 0.134 | 0.174 | |
| | G | 0.607 | 0.393 | 0.231 | 0.407 | 5.280 | 0.021** | 79 | 0.845 | 0.882 | |
| 9 | K | 0.574 | 0.426 | 0.756 | 0.766 | 0.019 | 0.889 | 92 | 0.999** | 0.999 | AC |
| | G | 0.598 | 0.402 | 0.189 | 0.317 | 3.222 | 0.072* | 63 | 0.036 | 0.052 | |

| | | | | | | | | | | | |
|---|---|---|---|---|---|---|---|---|---|---|---|
| 10 | K | 0.586 | 0.414 | 0.636 | 0.790 | 4.105 | 0.042** | 69 | 0.187 | 0.236 | |
| | G | 0.587 | 0.413 | 0.227 | 0.403 | 5.362 | 0.020** | 82 | 0.904 | 0.930 | AC |
| 11 | K | 0.432 | 0.568 | 0.831 | 0.812 | 0.090 | 0.764 | 73 | 0.358 | 0.422 | |
| | G | 0.700 | 0.300 | 0.162 | 0.222 | 0.776 | 0.378 | 70 | 0.858 | 0.897 | |
| 12 | K | 0.603 | 0.397 | 0.833 | 0.917 | 2.163 | 0.141 | 74 | 0.534 | 0.601 | |
| | G | 0.652 | 0.348 | 0.143 | 0.115 | 0.221 | 0.637 | 69 | 0.467 | 0.539 | |
| 13 | K | 0.591 | 0.409 | 0.798 | 0.803 | 0.006 | 0.933 | 82 | 0.915 | 0.939 | |
| | G | 0.611 | 0.389 | 0.174 | 0.241 | 1.011 | 0.314 | 89 | 0.996** | 0.998 | AC |
| 14 | K | 0.575 | 0.425 | 0.651 | 0.797 | 3.809 | 0.050* | 76 | 0.573 | 0.638 | AC |
| | G | 0.599 | 0.401 | 0.267 | 0.317 | 0.440 | 0.507 | 76 | 0.665 | 0.724 | |
| 15 | K | 0.579 | 0.421 | 0.862 | 0.952 | 3.311 | 0.068* | 72 | 0.332 | 0.395 | |
| | G | 0.602 | 0.398 | 0.133 | 0.050 | 2.777 | 0.095* | 87 | 0.978** | 0.989 | |

*Notes*: ** and * denote rejections at the 5% and 10% levels, respectively. AC denotes rejection of randomization under the Arellano–Carrasco test.

**Table 3.4.** Tests of Equality of Payoffs and Randomization Tests with Same Type of Player in Pairs

| | | Frequencies | | Win Rates | | Pearson Tests | | Runs Tests | | | |
|---|---|---|---|---|---|---|---|---|---|---|---|
| Pair # | Player | $L$ | $R$ | $L$ | $R$ | Statistic | $p$-value | $r$ | $\Phi[r-1,s]$ | $\Phi[r,s]$ | Arellano–Carrasco Tests |
| 1 | K | 0.575 | 0.425 | 0.767 | 0.828 | 0.824 | 0.363 | 76 | 0.573 | 0.638 | |
| | K | 0.520 | 0.480 | 0.179 | 0.236 | 0.732 | 0.392 | 69 | 0.112 | 0.147 | |
| 2 | K | 0.583 | 0.417 | 0.655 | 0.762 | 1.982 | 0.159 | 63 | 0.025 | 0.037* | AC |
| | K | 0.603 | 0.397 | 0.200 | 0.450 | 10.714 | 0.001** | 83 | 0.947 | 0.963 | AC |
| 3 | K | 0.574 | 0.426 | 0.756 | 0.672 | 1.282 | 0.257 | 68 | 0.124 | 0.162 | |
| | K | 0.607 | 0.393 | 0.264 | 0.305 | 0.303 | 0.581 | 79 | 0.845 | 0.882 | AC |
| 4 | K | 0.714 | 0.286 | 0.888 | 0.767 | 3.553 | 0.059* | 82 | 0.999** | 0.999 | AC |
| | K | 0.541 | 0.459 | 0.185 | 0.101 | 2.087 | 0.148 | 63 | 0.015 | 0.023** | AC |
| 5 | K | 0.777 | 0.223 | 0.479 | 0.636 | 2.563 | 0.109 | 69 | 0.999** | 0.999 | AC |
| | K | 0.557 | 0.443 | 0.429 | 0.561 | 2.579 | 0.108 | 82 | 0.863 | 0.896 | AC |
| 6 | K | 0.541 | 0.459 | 0.704 | 0.783 | 1.205 | 0.272 | 73 | 0.309 | 0.369 | |
| | K | 0.700 | 0.300 | 0.229 | 0.333 | 1.796 | 0.180 | 70 | 0.858 | 0.897 | AC |
| 7 | K | 0.572 | 0.428 | 0.884 | 0.922 | 0.593 | 0.441 | 74 | 0.440 | 0.507 | |
| | K | 0.636 | 0.364 | 0.074 | 0.145 | 1.993 | 0.157 | 69 | 0.351 | 0.418 | |
| 8 | K | 0.591 | 0.409 | 0.798 | 0.885 | 1.994 | 0.157 | 82 | 0.915 | 0.939 | AC |
| | K | 0.612 | 0.388 | 0.130 | 0.224 | 2.248 | 0.133 | 82 | 0.946 | 0.963 | AC |
| 9 | K | 0.575 | 0.425 | 0.721 | 0.828 | 2.356 | 0.124 | 76 | 0.573 | 0.638 | |
| | K | 0.609 | 0.391 | 0.143 | 0.373 | 10.586 | 0.001** | 76 | 0.691 | 0.749 | |

|  |  |  |  |  |  |  |  |  |  |  |  |
|---|---|---|---|---|---|---|---|---|---|---|---|
| 10 | K | 0.577 | 0.423 | 0.816 | 0.984 | 10.267 | 0.001** | 72 | 0.332 | 0.395 |  |
|  | K | 0.607 | 0.393 | 0.066 | 0.186 | 5.172 | 0.022** | 81 | 0.912 | 0.937 | AC |
| 11 | G | 0.640 | 0.360 | 0.927 | 0.889 | 0.636 | 0.424 | 69 | 0.386 | 0.456 |  |
|  | G | 0.612 | 0.388 | 0.054 | 0.138 | 3.139 | 0.076* | 54 | 0.000 | 0.001** | AC |
| 12 | G | 0.572 | 0.428 | 0.802 | 0.875 | 1.395 | 0.237 | 90 | 0.994** | 0.996 | AC |
|  | G | 0.612 | 0.388 | 0.120 | 0.241 | 3.800 | 0.051* | 70 | 0.323 | 0.388 |  |
| 13 | G | 0.575 | 0.425 | 0.756 | 0.703 | 0.520 | 0.470 | 65 | 0.048 | 0.068 |  |
|  | G | 0.616 | 0.384 | 0.293 | 0.224 | 0.874 | 0.349 | 91 | 0.999** | 0.999 | AC |
| 14 | G | 0.576 | 0.424 | 0.849 | 0.766 | 1.673 | 0.195 | 77 | 0.638 | 0.698 | AC |
|  | G | 0.566 | 0.434 | 0.247 | 0.108 | 4.712 | 0.029** | 82 | 0.872 | 0.904 |  |
| 15 | G | 0.586 | 0.414 | 0.932 | 0.887 | 0.919 | 0.337 | 78 | 0.737 | 0.789 |  |
|  | G | 0.611 | 0.389 | 0.109 | 0.052 | 1.458 | 0.227 | 78 | 0.822 | 0.863 | AC |

*Notes*: ** and * denote rejections at the 5% and 10% levels, respectively. AC denotes rejection of randomization under the Arellano–Carrasco test.

### LESSON 4

How to win in Rock–Scissors–Paper (RSP)? This is probably one of the first situations requiring mixed strategies that we humans encounter. To gain insights into the answer, we consider two recent studies.

Cook et al. (2011) studied a blindfolded player playing RSP against a sighted player, and their outcomes were compared to a control treatment in which two blindfolded players played. A tie was achieved almost exactly 1/3 of the time when the two blindfolded players met, but that rate increased to 36.3% in the blind-sighted treatment, a statistically significant difference. The authors attribute this difference to a subconscious tendency to imitate the actions of others. In particular, when the blind player completed his or her move more than 200 milliseconds before the sighted player, the sighted player had an increased tendency to play the same move.

Two hundred milliseconds is too fast for conscious reaction but still within the time necessary for the visual signal to be sent to the brain and an impulsive response signal to be sent to the hand. Now, 200 milliseconds or even just a few milliseconds is not too fast for the reaction of a top robot. If you can build such a robot, you can beat a human in RSP every single time. (Quick detour. Researchers at the Ishikawa Oku Lab at the University of Toyko recently built such an unbeatable robot with high-speed hands that works with a high-speed vision system. It takes only a single millisecond for this robot to recognize the shape of your hand, and just a few more milliseconds to form the shape that beats you. It all happens so fast that it is impossible to tell that the robot is waiting until you commit yourself to a move before it makes its move. Only the robot's ability to win 100% of the games might eventually give it away. Yep, score another one for the robots.)

Belot et al. (2013) extend the previous study by replacing the RSP game with a zero-sum Matching Pennies game, which creates far stronger incentives to avoid imitation for some subjects, with equally strong incentives to imitate for others. The results show that both intentional best-responding and automatic imitation occur.

Together with a large body of evidence of apparently spontaneous mimicry ("automatic imitation") in humans, these two articles supply the first direct evidence that both involuntary and intentional imitation occur in strategic games. Hence, they begin to illuminate the way that automatic and intentional processes interact in strategic experimental contexts. In particular, imitation raises novel issues concerning how strategic interactions are modeled in game theory and social sciences. These results also offer an important implication for the design of an experiment, such as the one in the previous chapter. In simultaneous games such as Matching Pennies and RSP, experimental subjects are formally required to present their gestures and make their choices simultaneously. But absolute simultaneity in every game is often difficult in practice. To remain as close as possible to the ideal model, one should then attempt to restrict as much as possible the possibility of imitation. This restriction can be easily accomplished in a laboratory with subjects seated separately at different computers. In face-to-face experiments with cards, seating subjects on opposite sides of an opaque partition board can prevent imitation. The absence of a partition, however, increases opportunities for automatic imitation (matching) and for intentionally avoiding imitation (mismatching), thereby transforming a simultaneous game into a sequential one. This difference means that the lack of a simple dividing board can influence and generate differences in behavior in strategic contexts via unintended changes in the game's structure. Be aware next time you implement or read about a face-to-face experiment that lacks a board.

# 4

## MAPPING MINIMAX IN THE BRAIN

(with Antonio Olivero, Sven Bestmann, Jose Florensa Vila, and Jose Apesteguia)

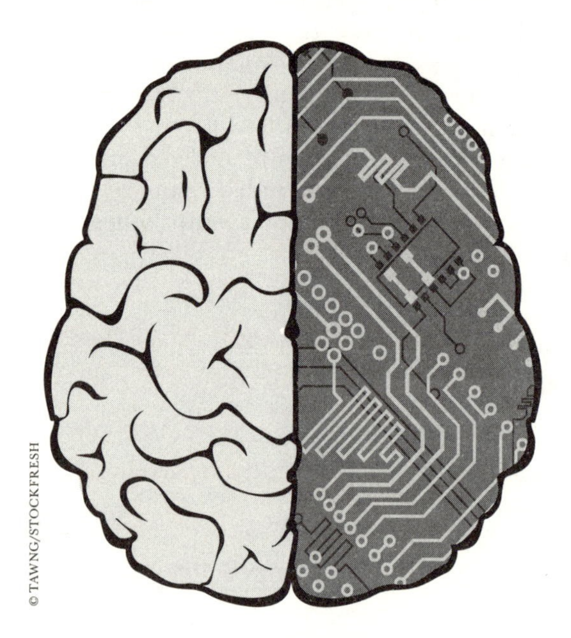



Soccer is a game you play with your brain.
—JOHAN CRUYFF, QUOTED IN ANDERSON AND SALLY 2013

Let there be granted to the science of pleasure what is granted to the science of energy; to imagine an ideally perfect instrument, a psychophysical machine, continually registering the height of pleasure experienced by an individual exactly according to the verdict of consciousness.
—FRANCIS Y. EDGEWORTH 1881

THE OBAMA ADMINISTRATION AND THE EUROPEAN COMMISSION ARE CURrently planning two different multiyear research efforts to produce an "activity map" that would show in unimaginable detail the working of the most complex organ in the body: the human brain. The first objective of these efforts is to gain a much deeper understanding of how the brain works. The final aim, probably unattainable for several decades, is to know how the brain generates perception, consciousness, memories, and thoughts and to find ways to intervene and influence such brain activities. Ideally, there will be many clinical benefits as well. The new knowledge could enable scientists to find better and cheaper ways to diagnose and treat depression, Parkinson's disease, stroke, schizophrenia, and other illnesses or injuries in the brain. Unfortunately, it is fair to say that we are a long way from such accurate understanding today.

Of the big scientific programs in the past century, few if any were as intimidating as these brain projects. The race between the United States and the USSR to put a man on the moon in the 1960s was relatively trivial because it was mostly achieved by technical processes, methods and knowledge that already existed at the time. The Human Genome Project to identify the complete sequence of genes on every chromosome in the body was completed a decade ago. There was no doubt that it was achievable; the only questions were when and at what cost. By contrast, the brain projects are not as clearly defined and will have to create new tools to explore the center of human cognition and behavior.

Thus far, "scientists have been able to infer the main function of certain regions of the human brain by studying patients with head injuries, brain tumors, and neurological diseases or by measuring oxygen levels and glucose consumption in the brains of healthy people" (Boffrey 2013).

In the past decade, this research has reached economics in what has become an entirely new field of scientific enterprise: neuroeconomics. This area combines mathematical frameworks, experimental methods, and lab and field behavioral data about peoples' choices with measures of neural activity. The goal is to relate formal theories of human choice to neural measures in an attempt to predict the effects of cognitive and emotional factors on individual choices.

Thus, using modern neuroimaging techniques—including functional magnetic resonance imaging (fMRI), positron emission tomography scans, and so on—economists have begun to look inside the brain and see what is going on when experimental subjects make economic decisions dealing with risk, uncertainty, gains, losses, endowments, temporal preferences, how to bid in auctions, and other behavior.[1]

"The data on say, dopamine release in the nucleus accumbens or bloody oxygen in the striatum when choosing between, say, an amount today or a bigger amount next month, are certainly fascinating in their own right. But can they improve our understanding of economic and social behavior?" (Maskin 2008). There is little agreement on this important question. Economists such as Colin Camerer, George Loewenstein, and

1 See, for instance, Camerer et al. (2005) and Glimcher et al. (2009).

Drazen Prelec (2005) predict that "we will eventually be able to replace the simple mathematical ideas that have been used in economics with more neurally-detailed descriptions." By contrast, economic theorists Faruk Gul and Wolfgang Pesendorfer (2008) argue that neuroscience evidence is irrelevant to economics because "the latter makes no assumptions and draws no conclusions about the physiology of the brain."[2]

In principle, the Gul–Pesendorfer critique would seem to be right, at least if we limit ourselves to current practice in economics. Basically, in a standard economic model, a person is presented with several options, and we try to predict which one he or she will choose. There is no need to know or infer anything about his or her brain as long as the prediction is correct. The problem is that predictions are sometimes far from correct, and so, in principle, we might improve the model by allowing behavior to depend not only on the economic options but also on sensual and neurological data about the person. From this perspective, neural findings show great potential for improving economic analysis. One can be cautiously optimistic and think that exploring new technologies and getting new data has option value.

With this objective in mind, this chapter is concerned with mixed strategies. Using fMRI techniques, we peer inside the brain when experimental subjects play the penalty kick game. As we have noted already, minimax is considered a cornerstone of interactive decision-making analysis. More importantly, to the best of our knowledge, the minimax strategies have not been mapped in the brain previously by studying simultaneously the two testable implications of equilibrium.

## SUBJECTS

The study was performed in the Hospital Nacional de Parapléjicos de Toledo (Spain) during 2012 with 40 healthy subjects. They formed 20 pairs. Twenty volunteers were studied in the MRI (6 women, mean age 30.7±6.0 (SD) years, range 21–40 years). Six subjects had low-medium education level (primary or high school), and 14 had high educational level (college degree or higher). Twenty subjects were studied outside the scanner (6 women, mean age 33.0±7.6 (SD) years, range 21–44 years). Six subjects had low-medium education level (primary or high school), and 14 had high educational level (college degree or higher). All subjects gave informed consent before participation. The study was approved by the local ethics committee and was conducted in accordance with the World Medical Association's Declaration of Helsinki.

2 See, however, the pioneering recent work by Brocas and Carrillo (2008) and Alonso et al. (2013).

## EXPERIMENTAL SETUP

Two players were engaged in the penalty kick game. They were not friends and had not met before; recall that this aspect is important. One player was playing in a computer in a quiet room located outside the scanner room. The other player was lying down within the MRI room. For this player, the PC monitor was substituted by MRI-compatible goggles and the keyboard was substituted by a button box designed for the hand. Player location (inside or outside the MRI) was decided by flipping a coin. Both PC screen and goggles were always displaying the matrix in the middle and upper part of the screen that was studied in the previous chapters:

| | *A* | *B* |
|---|---|---|
| *A* | 60 | 95 |
| *B* | 90 | 70 |

In the lower part of the screen, the subjects saw a few lines of text where they received the "go" signal to make their decisions and where they also received the feedback about the opponent's decision and the identity of the winner. The prize was 1 euro per trial, and the subjects performed 100 to 120 trials (with the exception of one pair that performed 82 trials). Both players always received simultaneously the go signal and the feedback after each round (choices and outcomes).[3]

Table 4.1 collects the results of the Pearson tests of equality of payoffs across *A* and *B* strategies, as well as the runs tests, for each pair. As may be observed, quite remarkably, subjects play rather close to minimax. The subjects outside the MRI play in fact basically according to the equilibrium hypothesis with just two rejections in the Pearson tests and three in the runs tests at both the 5% and 10% levels. Perhaps even more remarkable is that 70% of the subjects within the MRI passed the Pearson test and 60% passed the runs test, something that is an order of magnitude closer to minimax than MLS players or friends.

## SCANNING AND IMAGE PROCESSING

Some technicalities may be standard in neurological research but are probably unfamiliar to most economists. Each scanning session on a 3T Siemens Trio system comprised functional T2*-weighted MRI

3 The instructions and other details of the experimental procedure are available in Palacios-Huerta et al. (2013). They are almost identical to those in chapter 2, except for the aspects that concerned the fMRI that were included in the experimental instructions.

**Table 4.1.** Pearson and Runs Tests in the Penalty Kick fMRI Experimental Game

| Pair | Player | $N$ | $L$ | $R$ | $Ls$ | $Lf$ | $Rs$ | $Rf$ | $p$ | $\chi^2$ stat | $p$-value | $r$ | $\Phi[r-1,s]$ | $\Phi[r,s]$ |
|---|---|---|---|---|---|---|---|---|---|---|---|---|---|---|
| 1 | MRI | 82 | 37 | 45 | 30 | 7 | 34 | 11 | 0.78 | 0.36 | 0.547 | 41 | 0.401 | 0.490 |
| | Out | 82 | 42 | 40 | 32 | 10 | 32 | 8 | 0.78 | 0.17 | 0.677 | 38 | 0.159 | 0.219 |
| 2 | MRI | 103 | 72 | 31 | 57 | 15 | 26 | 5 | 0.81 | 0.31 | 0.580 | 33 | 0.002 | 0.005** |
| | Out | 103 | 46 | 57 | 36 | 10 | 47 | 10 | 0.81 | 0.29 | 0.593 | 57 | 0.820 | 0.868 |
| 3 | MRI | 119 | 88 | 31 | 70 | 18 | 25 | 6 | 0.80 | 0.02 | 0.896 | 35 | 0.001 | 0.003** |
| | Out | 119 | 52 | 67 | 44 | 8 | 51 | 16 | 0.80 | 1.31 | 0.252 | 60 | 0.495 | 0.570 |
| 4 | MRI | 117 | 74 | 43 | 61 | 13 | 36 | 7 | 0.83 | 0.03 | 0.858 | 61 | 0.846 | 0.888 |
| | Out | 117 | 58 | 59 | 48 | 10 | 49 | 10 | 0.83 | 0.00 | 0.967 | 83 | 0.999** | 0.999 |
| 5 | MRI | 116 | 66 | 50 | 60 | 6 | 38 | 12 | 0.84 | 4.82 | 0.028** | 45 | 0.005 | 0.009** |
| | Out | 116 | 73 | 43 | 60 | 13 | 38 | 5 | 0.84 | 0.79 | 0.375 | 20 | 0.000 | 0.000** |
| 6 | MRI | 119 | 69 | 50 | 54 | 15 | 43 | 7 | 0.82 | 1.15 | 0.283 | 54 | 0.150 | 0.198 |
| | Out | 119 | 46 | 73 | 7 | 39 | 15 | 58 | 0.18 | 0.53 | 0.466 | 59 | 0.581 | 0.655 |
| 7 | MRI | 116 | 63 | 53 | 53 | 10 | 40 | 13 | 0.80 | 1.36 | 0.244 | 68 | 0.953 | 0.968 |
| | Out | 116 | 65 | 51 | 51 | 14 | 42 | 9 | 0.80 | 0.27 | 0.602 | 47 | 0.013 | 0.021** |
| 8 | MRI | 116 | 82 | 34 | 70 | 12 | 23 | 11 | 0.80 | 4.75 | 0.029** | 40 | 0.015 | 0.026* |
| | Out | 116 | 72 | 44 | 58 | 14 | 35 | 9 | 0.80 | 0.02 | 0.895 | 52 | 0.207 | 0.268 |
| 9 | MRI | 118 | 67 | 51 | 52 | 15 | 39 | 12 | 0.77 | 0.02 | 0.884 | 51 | 0.056 | 0.081 |
| | Out | 118 | 62 | 56 | 45 | 17 | 46 | 10 | 0.77 | 1.52 | 0.217 | 68 | 0.922 | 0.945 |
| 10 | MRI | 119 | 62 | 57 | 54 | 8 | 43 | 14 | 0.82 | 2.68 | 0.102 | 61 | 0.507 | 0.580 |
| | Out | 119 | 73 | 46 | 55 | 18 | 42 | 4 | 0.82 | 4.77 | 0.029** | 51 | 0.088 | 0.124 |

|  |  |  |  |  |  |  |  |  |  |  |  |  |  |  |
|---|---|---|---|---|---|---|---|---|---|---|---|---|---|---|
| 11 | MRI | 100 | 64 | 36 | 57 | 7 | 26 | 10 | 0.83 | 4.63 | 0.031** | 27 | 0.000 | 0.000** |
|  | Out | 100 | 62 | 38 | 51 | 11 | 32 | 6 | 0.83 | 0.06 | 0.801 | 49 | 0.532 | 0.615 |
| 12 | MRI | 105 | 61 | 44 | 53 | 8 | 35 | 9 | 0.84 | 1.01 | 0.314 | 68 | 0.999** | 0.999 |
|  | Out | 105 | 54 | 51 | 41 | 13 | 47 | 4 | 0.84 | 5.09 | 0.024** | 48 | 0.121 | 0.165 |
| 13 | MRI | 113 | 75 | 38 | 57 | 18 | 27 | 11 | 0.74 | 0.32 | 0.569 | 47 | 0.147 | 0.201 |
|  | Out | 113 | 49 | 64 | 14 | 35 | 15 | 49 | 0.26 | 0.38 | 0.536 | 56 | 0.423 | 0.499 |
| 14 | MRI | 114 | 78 | 36 | 67 | 11 | 20 | 16 | 0.76 | 12.55 | 0.000** | 41 | 0.016 | 0.028* |
|  | Out | 114 | 72 | 42 | 17 | 55 | 10 | 32 | 0.24 | 0.00 | 0.981 | 61 | 0.903 | 0.934 |
| 15 | MRI | 104 | 47 | 57 | 39 | 8 | 38 | 19 | 0.74 | 3.57 | 0.059* | 38 | 0.001 | 0.002** |
|  | Out | 104 | 59 | 45 | 18 | 41 | 9 | 36 | 0.26 | 1.47 | 0.226 | 45 | 0.064 | 0.094 |
| 16 | MRI | 115 | 62 | 53 | 52 | 10 | 40 | 13 | 0.80 | 1.26 | 0.262 | 58 | 0.451 | 0.526 |
|  | Out | 115 | 56 | 59 | 14 | 42 | 9 | 50 | 0.20 | 1.71 | 0.192 | 57 | 0.356 | 0.428 |
| 17 | MRI | 114 | 83 | 31 | 74 | 9 | 23 | 8 | 0.85 | 3.98 | 0.046** | 43 | 0.192 | 0.264 |
|  | Out | 114 | 57 | 57 | 8 | 49 | 9 | 48 | 0.15 | 0.07 | 0.793 | 57 | 0.388 | 0.462 |
| 18 | MRI | 117 | 81 | 36 | 62 | 19 | 29 | 7 | 0.78 | 0.23 | 0.630 | 48 | 0.232 | 0.304 |
|  | Out | 117 | 48 | 69 | 8 | 40 | 18 | 51 | 0.22 | 1.45 | 0.228 | 52 | 0.120 | 0.163 |
| 19 | MRI | 106 | 55 | 51 | 41 | 14 | 34 | 17 | 0.71 | 0.79 | 0.373 | 55 | 0.544 | 0.620 |
|  | Out | 106 | 53 | 53 | 19 | 34 | 12 | 41 | 0.29 | 2.23 | 0.135 | 52 | 0.312 | 0.384 |
| 20 | MRI | 119 | 69 | 50 | 54 | 15 | 43 | 7 | 0.82 | 1.15 | 0.283 | 54 | 0.150 | 0.198 |
|  | Out | 119 | 46 | 73 | 7 | 39 | 15 | 58 | 0.18 | 0.53 | 0.466 | 59 | 0.581 | 0.655 |

*Notes*: The columns *Ls*, *Lf*, *Rs*, and *Rf* denote successes (*s*) and failures (*f*) when choosing *L* and *R*, respectively. The variable $p$ denotes the success rate obtained in the experiment and ** and * indicate rejection at the 5% and 10% levels, respectively.

transverse EPIs with blood oxygenation level-dependent (BOLD) contrast (40 slices per volume, TE: 61 ms; TR: 2.43 s; 3 × 3 mm in-plane resolution; 3-mm slice thickness), and one experimental session with 950 volumes was acquired for each participant. Additionally, a whole-head T1-weighted anatomical image was acquired after the experiment using a standard FLASH sequence with an isotropic resolution of 1 $mm^3$. Imaging data were analyzed using Statistical Parametric Mapping (SPM5, http://www.fil.ion.ucl.ac.uk/spm) implemented in MATLAB 10. The first five volumes were discarded for T1-signal equilibration effects. All remaining volumes were realigned to the first volume to correct for interscan head movements. Additionally, interactions of head motion and geometric distortions were removed using the "unwarp" toolbox as implemented in SPM5 (Andersson et al. 2001). The EPI images were normalized to a standard EPI template based on the Montreal Neurological Institute (MNI) reference brain in Talairach space. An AR(1) model accounted for serial autocorrelations of the data, and spatial smoothing of normalized images with an isotropic 8-mm full-width at half-maximum Gaussian kernel was conducted to allow for valid statistical inference according to Gaussian random field theory.

## IMAGING ANALYSES

Here are more technicalities. Single-subject fixed-effects models were computed for each participant by multiple regression of the voxelwise time series onto a composite model containing the covariates of interest. These included the decision epoch, choice display epoch, and outcome epoch. Additionally, response key presses were included and modeled as delta functions. All covariates were convolved with a canonical synthetic hemodynamic response function in a general linear model (Friston et al. 1995, 1998) together with a single covariate representing the mean (constant) term over scans. Voxelwise parameter estimates for each covariate were calculated, resulting from the weighted least squares fit of the model to the data.

At a second (group) level of analysis, the contrast images for each participant and covariate were submitted to a 1-sample *t*-test for each covariate of interest in a random-effects analysis across participants.

## GROUP ACTIVATION RESULTS

The main results revealed activity increases in various bilateral prefrontal regions during the decision period. Interestingly, the data analysis suggested that activity in the left inferior prefrontal cortex related significantly to the ability to equate payoffs (as measured by the *p*-value), one of two key criteria for successfully playing the game (see figure 4.1). In other words, across the group, activity in this prefrontal region correlated with the performance measure for equating payoffs, with higher activity in participants who more effectively succeeded in equating payoffs. Conversely, a contralateral right inferior prefrontal region related to the ability to generate random sequences of choices (see figure 4.2). Activity in these regions was correlated with the performance score testing for the randomness of choices using the *p*-value of the runs test.

Together these pilot data suggest that two inferior prefrontal nodes jointly contribute to the ability to optimally play our asymmetric zero-sum penalty kick game by ensuring the appropriate equating of payoffs across strategies and the generating of random choices within the game, respectively.

This evidence, therefore, contributes to the neurophysiological literature studying competitive games. Vickery and Jiang (2009), for instance, find that the right inferior parietal lobule was systematically activated in the course of the play of a classic Matching Pennies game. In Hampton et al. (2008), models are built on the basis of various behavioral assumptions (such as fictitious play, reinforcement learning, or a formulation of the influence of one's actions on the others) that describe mentalizing in a version of a Matching Pennies game. They find that the medial prefrontal cortex (mPFC) and posterior superior temporal sulcus (pSTS) were activated.[4] Seo et al. (2009) record neural firing rates directly, and they show that firing rates are consistent with reinforcement learning in the Matching Pennies game. With respect to randomization per se (that is, not in the context of a strategic interaction), there is sound evidence using a variety of techniques that the dorsolateral prefrontal cortex is activated in the process of generating random sequences of numbers (see, e.g., Jahanshahi et al. 1998 and Daniels et al. 2003). In Ischebeck et al. (2008), the random generation of items from an ordered structure, such as numbers, activates the intraparietal sulcus more intensively than when using items from a nonordered structure, such as different animals.

The long-run goal of neuroeconomics is "to create a theory of economic choice and exchange that is neurally detailed, mathematically accurate, and behaviorally relevant" (Camerer 2008). This chapter contributes in this direction by combining the classic mathematical framework of strictly competitive games, experimental methods, and lab data on a strategic situation that is considered a cornerstone of interactive decision-making (minimax) and providing measures of neural activity in the two dimensions that characterize the equilibrium.

4 See also Kadota et al. (2010) and Vickery et al. (2011).

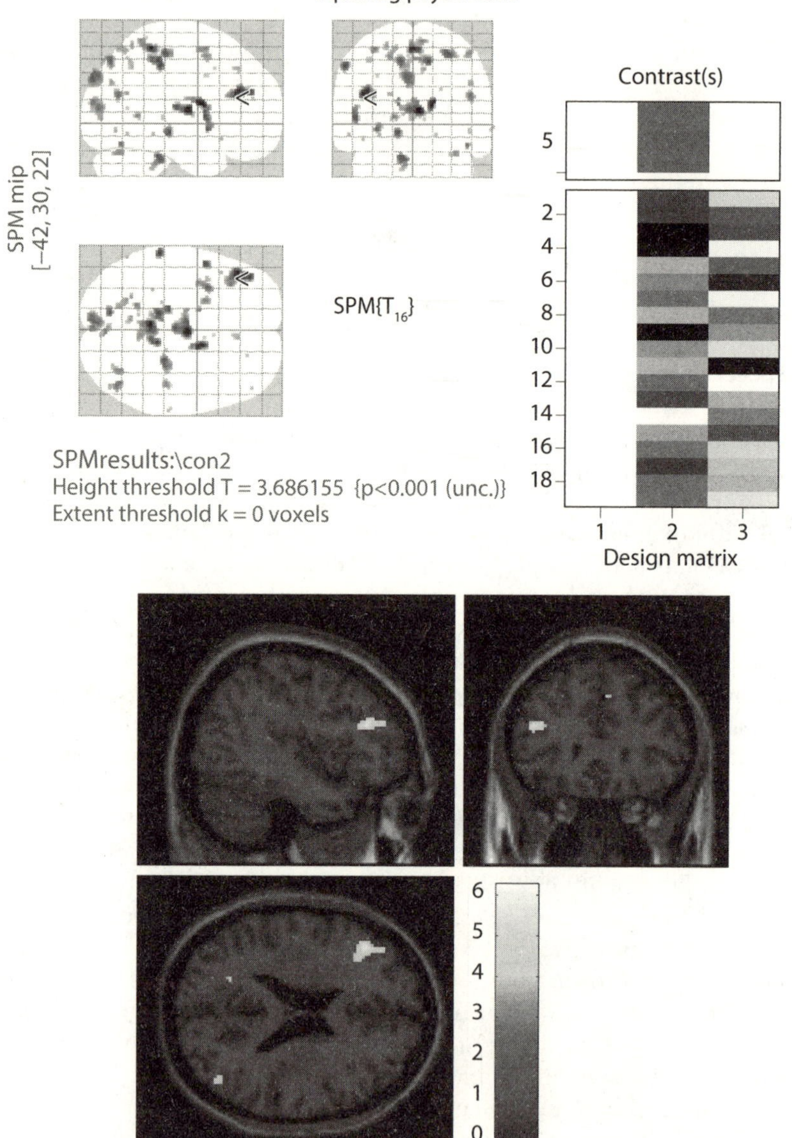


**Figure 4.1.** Brain activity when subjects equate payoffs across strategies.

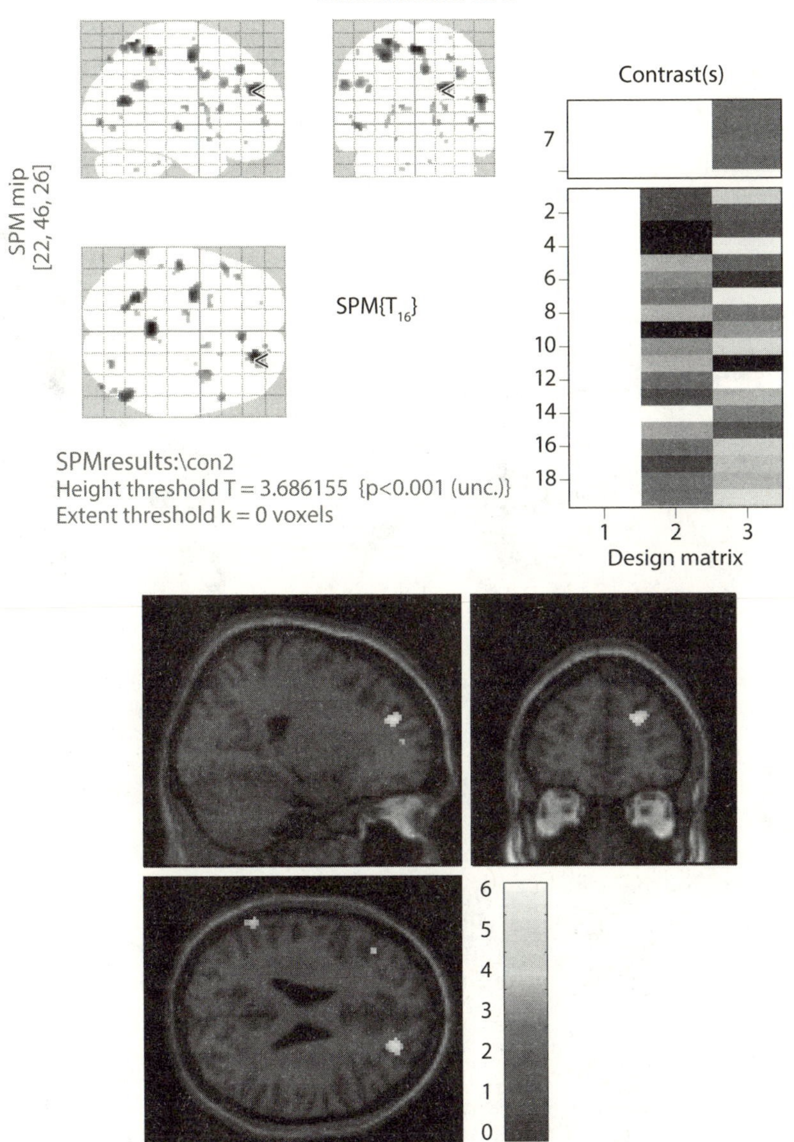


**Figure 4.2.** Brain activity when subjects randomize their strategies.

5

# PSYCHOLOGICAL PRESSURE ON THE FIELD AND ELSEWHERE

© TAWNG/STOCKFRESH

The secret of getting ahead is getting started.
—Attributed to Mark Twain (1835–1910)

In the early 1960s, Rafael Ballester was a prestigious journalist in Cádiz, a city in the south of Spain. Cádiz is well known in the world of football because for decades it has organized a famous international summer tournament in early August, the Trofeo Ramón de Carranza. Each year, four different teams are invited. They play four games in two days, the semifinals on Saturday and, on Sunday, the match that determines the third and fourth place in the tournament is played, followed by the final.

Quite often the semifinals on Saturday ended up being tied, and the teams had to play for extra time. If they remained tied after the extra time, then, in keeping with tradition, a coin would decide the team that would play in the final the following day. This was the prevailing system in FIFA to break ties up to 1970. It was problematic. Teams would often be quite tired after the additional effort from the previous day and, with no rest days, the quality of soccer would suffer substantially on Sunday. Plus, there was, of course, the added unpleasant feature that many times it was the arbitrariness of a coin toss that decided the outcome of a match, sometimes even the final winner of the tournament.

Mr. Ballester had an ingenious idea to overcome these problems: ties would be resolved with a *penalty shoot-out* where both teams would kick the same number of penalty kicks until one has scored more goals than the other could score. For instance, both teams could begin by kicking five penalty kicks each. If they remained tied, they would kick more, until a winner was declared.

He published his idea in the newspaper *Diario de Cádiz* in August 1957, right after the final between Athletic Club de Bilbao and Sevilla that was decided by a coin.[1] The organizers liked it and decided to adopt it. The first opportunity to put the new system into action was on September 2, 1962, in the final match of the tournament between Barcelona and Zaragoza. Barcelona won after six penalty kicks. This procedure to break ties quickly gained popularity and spread to several friendly tournaments in Europe, Africa, and South America in following years. Soccer connoisseurs may remember, for instance, the Trofeo Corpus Christi from 1964, played in Orense (Spain), which featured three teams: FC Porto from Portugal, RC Deportivo La Coruña from Spain, and Athletic Club de Bilbao. The first game between RC Deportivo La Coruña and Athletic Club de Bilbao ended 1–1. In the penalty shoot-out, first Deportivo kicked five penalty kicks in one go. José Ángel Iribar, one of the best goalkeepers in soccer history, stopped four of them (an incredible performance since around 80% of penalties are scored on average), and the fifth one hit the goalpost. Then Athletic Club de Bilbao scored just its first penalty kick and won the match.

The popularity of the new system to break ties was such that, in 1970, FIFA decided to adopt it. There are no detailed minutes of the International Board Meeting held on June 27, 1970, at the Caledonian Hotel in Inverness, Scotland, when the shoot-out proposal was approved, but the idea of one team taking all penalties in one go was replaced by the system of alternate penalties that we know today.

Beginning on that date, the method of determining the winning team, where competition rules require that one team is declared the winner after a drawn match, was by a penalty shoot-out. Although it was too late for the Mexico World Cup in 1970, this decision meant that it would be used worldwide in all the major elimination tournaments involving both national teams (e.g., World Cups, European Cups, American Cups) and club teams (e.g., Champions League, UEFA Cup) from then on.

1 The story is reported in Relaño (2010). Some sources mistakenly credit Israeli Yosef Dagan as the inventor of the penalty shoot-out. After watching the Israeli team lose an Olympic quarter-final by drawing of lots in 1968, he proposed this system in a letter to the Israel Football Association. Others credit former German referee Karl Wald in a proposal to the Bavarian Football Association in 1970.

The basic rules of a penalty shoot-out were as follows:

- The referee tosses a coin and the team whose captain wins the toss takes the first kick.
- The referee keeps a record of the kicks being taken.
- Subject to the conditions explained below, both teams take five kicks.
- The kicks are taken alternately by the teams.
- If, before both teams have taken five kicks, one has scored more goals than the other could score, even if it were to complete its five kicks, no more kicks are taken.
- If, after both teams have taken five kicks, both have scored the same number of goals, or have not scored any goals, kicks continue to be taken in the same order until one team has scored a goal more than the other from the same number of kicks.

History says that the first penalty shoot-out in a senior official competitive football match took place in England on August 5, 1970, just a few days after the FIFA approval. Manchester United was the first winner, defeating Hull City 4–3 on penalties in the semifinal of the Watney Cup. The set of five players from the first team that kicked in a penalty shoot-out included some of the greatest players ever: George Best, Brian Kidd, Bobby Charlton, Denis Law, and Willie Morgan, and Alex Stepney was in goal. First trivia alert: The first player to score in a shoot-out was George Best, with a low shot to the keeper's right. Second trivia alert: The first player to miss a kick in a shoot-out was Denis Law. Hull's keeper, McKechnie, dived to his right to save it. And a trivia question: What was the world's longest penalty shoot-out? The answer is in chapter 10.

This system was in place until July 2003, when FIFA decided to change the first regulation in the procedure slightly by replacing it with the following:

- The referee tosses a coin, and the team whose captain wins the toss *decides* whether to take the first or the second kick.

The clarity of the rules of a penalty shoot-out, as well as the characteristics and the detailed structure of a penalty kick discussed in the previous chapters, offer substantial advantages for studying the role that psychological elements (emotions) may play in dynamic competitive environments. As Miller (1998) notes, right from the beginning, the *Daily Telegraph* confirmed the presence of emotional elements in this setting. After the Manchester United–Hull final, it wrote: "This was the first time this method of settling a match had been used at senior level in England and it must be rated a resounding success. The suspense, as five players from each side fired alternately, was almost intolerable." The

*Daily Mail* said, "The penalty-taking session which settled this pulsating game was one of the most exciting and dramatic features I have ever seen on a soccer field."

At least since Hume (1739) and Smith (1759), psychological elements have been argued to be as much a part of human nature, and possibly as important for understanding human behavior, as the strict rationality considerations included in economic models that adhere to the rationality paradigm. This idea suggests that any rational theory of human behavior that omits these elements may yield results of unknown reliability until confronted with the data.

Motivated by evidence from new and richer data sets during the past couple of decades, an important body of research has attempted to parsimoniously incorporate psychological motives into standard economic models. The empirical evidence to test these models is typically obtained from the observation of human decision-making in laboratory environments, where experiments have the important advantage of providing control over relevant margins. A great deal of laboratory evidence has been accumulated demonstrating circumstances under which strict rationality considerations break down and other patterns of behavior, including psychological considerations, emerge. Nature, however, is less willing to contribute with empirical evidence. In fact, it rarely creates the circumstances that allow a clear view of the psychological principles at work. And when it does, the phenomena are typically too complex to be empirically tractable in a way that allows psychological elements to be discerned within the characteristically complex behavior exhibited by humans.[2]

This is why a penalty shoot-out is important. It provides an unusually clean opportunity in a real-world environment to discern the presence of psychological elements. In addition to the virtues of a penalty kick described in previous chapters,

1. A penalty shoot-out is a randomized natural experiment, that is, a real-life situation in which the treatment and control groups are determined via explicit randomization. In this case, the treatment that is randomly given to one team is the *order* of play: One team goes first in the sequence of tasks (penalty kicks) and the other second. As is well known, randomized experiments provide researchers with the critical advantage that they guarantee that the conditions for causal inference are satisfied.
2. The subjects involved in a shoot-out are professionals who have to perform a simple task: kick a ball once.

2 See Rabin (1998) and DellaVigna (2009) for excellent surveys of existing work.

3. All the relevant variables that are typically hard to observe and measure in other settings can be observed and measured.
4. And, finally, the analysis of a penalty shoot-out is also important scientifically because it relates to several strands of literature in economics and psychology:

    a. First, the natural setting corresponds to what is known as a tournament. Tournament competitions are pervasive in organizations and in real life, and often characterize situations such as competitions for promotion in internal labor markets in firms and organizations, patent races, political elections, and many others. As a framework of analysis, the tournament model was formally introduced by Lazear and Rosen (1981), and over the past couple of decades a large literature has studied a number of important aspects of this incentive scheme both theoretically and empirically.[3]

    Despite the large body of work, however, there is very little evidence documenting how psychological or emotional effects may be relevant in explaining the performance of subjects competing in tournament settings. Difficulties in clearly observing actions, outcomes, choices of risky strategies, and other relevant variables in a real-life tournament are often exceedingly high, and as a result it is typically impossible to discern the extent to which psychological elements may explain performance with sufficient precision.

    The characteristics of a penalty shoot-out, however, are ideal for overcoming these obstacles. Variables such as the choice of effort levels and risky strategies that are typically hard to observe and measure play no role in this setting: The task (kicking a ball once) involves little physical effort and, with only two possible outcomes (score or no score), risk plays no role either.[4] Outcomes (goal or no goal) can be perfectly observed and are immediately determined after players make their choices. The fact that there is no subsequent play and that the task is immediate (a penalty kick takes less than half a second) is indeed critical to cleanly interpreting the empirical evidence.

3 See Nalebuff and Stiglitz (1983) and Rosen (1986) for early contributions, and Prendergast (1999) for a review.

4 The role of risk in tournament competitions has been studied in Hvide (2002) and Hvide and Kristiansen (2003). In dynamic competition games, there is a literature on the "increasing dominance" effect of a leader over a rival, which studies the strategic amount of resources to use and allocate throughout a competition (Cabral 2003).

b. Second, an important literature in social psychology has studied expert performance and performance under pressure such as that induced by high stakes, the presence of an audience, and other aspects.[5] In a penalty shoot-out, however, both teams have the same stakes and both perform in front of the same audience. The explicit randomization procedure that is used to determine the kicking order means that there is no reason why one team should be systematically more affected than the other team by the stakes or the audience. The coin does not know which team is supported by the home audience (if any) or has greater stakes. What is new from the perspective of the existing academic literature is that differences in the interim state of the competition caused by the randomly determined kicking order may generate differences in psychological elements that could have an effect on performance.[6]

c. Finally, there is some economic literature on the ex post fairness of certain regulations in sports where a coin flip that determines the order of play may have a significant effect on the outcome of a game by giving the winner of the coin flip *more* chances to perform a task (see, for example, Che and Hendershott (2008) for the case of extra-time sudden-death regulations in the US National Football League). In a penalty shoot-out setting, however, we are under ideal circumstances: A coin flip determines only the order of competition, and both teams have exactly the *same* chances to perform a task. Yet, human nature may be such that the outcome of a perfect randomized trial has to be considered ex post unfair if in fact the order is shown to matter for performance for psychological reasons.

We take data from the Union of European Football Associations (UEFA), the Rec.Sport.Soccer Statistics Foundation, the Association of Football Statisticians in the United Kingdom, the Spanish newspapers *Marca* and *El Mundo Deportivo,* www.weltfussball.de, and the archives of various soccer clubs. The data set comprises 1,001 penalty shoot-outs with 10,431 penalty kicks over the period 1970–2013. It is comprehensive in that it includes virtually all the penalty shoot-outs in the history of the

5 See, for instance, Ericsson et al. (2006) and Beilock (2010). Ariely et al. (2009) review and discuss this literature.

6 In contrast to the size of the psychology literature, the economics literature is fairly limited, with pioneering theoretical contributions by Loewenstein (1987), Caplin and Leahy (2001), and Rauh and Seccia (2006) on anxiety and anticipatory emotions. There are, however, no previous empirical contributions with evidence from strictly competitive environments in real life.

main international elimination tournaments involving national teams (e.g., the World Cup, European Championship, and American Cup) and club teams such as UEFA Champions League and the UEFA Cup (now known as the Europa League). It also includes data on national club elimination tournaments such as the Spanish Cup, German Cup, and the English Football Association Cup.

This chapter follows Apesteguia and Palacios-Huerta (2010), AP henceforth, and for every shoot-out of every competition, it collects information on the date, the identity of the teams kicking first and second, the final outcome of the shoot-out, the outcomes of each of the kicks in the sequence, the geographical location of the game (that is, whether the game was played in a home ground, a visiting ground, or in a neutral field) and variables that measure the quality of the teams, such as their previous experience in shoot-outs, their official FIFA and UEFA rankings (for national teams), and the division, category, and standings (for club teams).[7]

As is well known, and following the description in Manski (1995), let $y_z$ be the outcome that a subject (a team in our case) would realize if he or she were to receive treatment $z$, where $z=0,1$. Let $P(y_z|x)$ denote the distribution of outcomes that would be realized if all subjects with covariates $x$ were to receive treatment $z$. The objective is to compare the distributions $P(y_1|x)$ and $P(y_0|x)$. When the treatment $z$ received by each subject with covariates $x$ is statistically independent of the subject's outcomes, we have $P(y_z|x)=P(y_z|x, z=1)=P(y_z|x, z=0)$ for $z=0,1$. Now let $y \equiv y_1 z + y_0(1-z)$ denote the outcome actually realized by a member of the population, namely, $y_1$ when $z=1$ and $y_0$ when $z=0$. Note that $P(y|x, z=1)=P(y_1|x, z=1)$ and $P(y|x, z=0)=P(y_0|x, z=0)$. Hence, if we denote by $B$ the specified set of outcome values (that is, simply win or lose in our case), when the treatment is independent of outcomes, the estimate of the treatment effect $T(B|x)$ is simply the following:

$$T(B|x) = P(y \in B|x, z = 1) - P(y \in B|x, z = 0)$$

Next, we extend the analysis in AP (2010) by studying not only the data for 1970–2003 but also the data for the following decade as well, that is, 43 years: 1970–2013. Note that the average treatment effect is identical before and after 2003. The fact that after 2003 players are required to choose the order (whether to kick first or second) is irrelevant for the

7 Consistent with the randomization procedure used to determine the order of play, it is not possible to reject the null hypothesis that any of these characteristics are irrelevant in determining the order of play, at the usual levels of significance. That is, the coin does not systematically select a specific type of teams with certain characteristics to kick first or second.

size of the average treatment effect. Their choices are interesting as a test of rationality, or consistency, but it does not affect $T(B|x)$.

To see this effect, consider a shoot-out between teams $i$ and $j$ in the framework of Bhaskar (2009). Let $w$ denote the state of the world that captures all relevant factors including the characteristics of the two teams, and let $p(w)$ be the win probability for $i$ when $i$ shoots first, and $q(w)$ the win probability when it shoots second. Under random assignment of the treatment "shooting first" (period 1970–2003), the probability that the team that shoots first wins is given by $0.5\{p(w) + [1 - q(w)]\} = 0.5[1 + \lambda(w)]$ where $\lambda(w) = p(w) - q(w)$. Obviously, $\lambda(w)$ can be negative for some $w$. Let us call this number $E(\lambda)$. Consider now the period after 2003, where the winner of the coin toss chooses the order. If the players always choose optimally, then the win probability for the team kicking first is exactly identical to $E(\lambda)$. But consider the opposite scenario: The winner of the coin toss always makes the inferior choice, that is, the winner chooses first when it should choose second, and second when it should choose first. Then the estimated treatment effect is also *exactly equal* to $E(\lambda)$. And the same, of course, in any intermediate scenario where the winner sometimes chooses to kick first and others second. All we can conclude after 2003 is the rationality or irrationality (the correctness or incorrectness) of the choices the teams make; the average treatment effect remains unchanged.[8]

Figure 5.1 and table 5.1 report $T(B)$ unconditional on any variables. The data show that kicking first conveys a strongly significant (beyond the 1% level) and sizable advantage: The team that kicks first wins the penalty shoot-out around 60% of the time.

Thus, the data show that a penalty shoot-out is not a 50–50 lottery. It is more like a 60–40 lottery where the first-kicking team has 20% more tickets. As expected, using a regression framework to provide an estimate of the treatment effect conditional on the complete set of available characteristics for the teams under various probit and logit specifications yields the same results. The order of play is strongly significant in every specification in table 5.2, and there is a significant and sizable advantage to the team that is first to kick. Mapping the regression coefficient into the corresponding Normal and Logistic distribution yields an effect in the most complete specifications of columns two and four in this table, again, slightly above 60% for the team that kicks first.

What the clean natural experiment just studied allows us to identify is that the nature of the mechanism generating these differences in performance is psychological. These emotional effects are endogenous to

8 Bhaskar (2009) offers a more detailed analysis, with an excellent application to the consistency of batting choices in cricket.

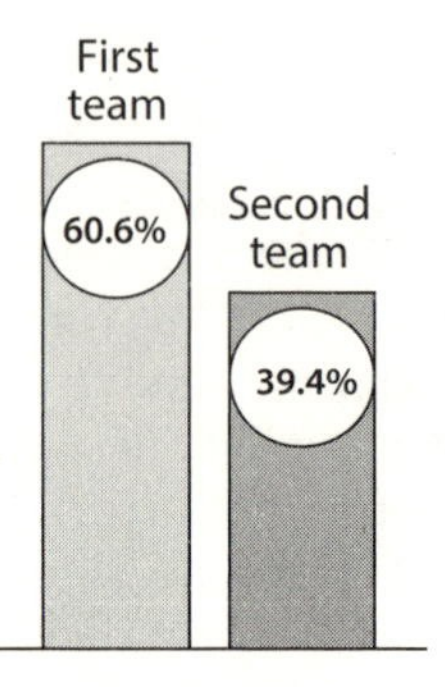


**Figure 5.1.** Winning frequencies by team, 1970–2013.

**Table 5.1.** Percentage First Team Wins in International and National Competitions 1970–2012

| | Number of shoot-outs | First team wins (%) | |
|---|---|---|---|
| International Competitions | | | |
| 1. National Teams | | | |
| World Cup | 22 | 59.1% | |
| European Championship | 15 | 33.3% | |
| Copa América | 18 | 61.1% | |
| African Nations Cup | 20 | 60.0% | |
| Gold Cup | 10 | 70.0% | |
| Asian Cup | 16 | 56.3% | |
| 2. Club Teams | | | |
| European Champions League | 49 | 63.3% | |
| European Cup Winners' Cup | 32 | 62.5% | |
| UEFA Cup | 110 | 55.5% | |
| National Competitions | | | |
| German Cups | 183 | 49.7% | |
| English Cups | 179 | 53.6% | |
| Spanish Cup | 347 | 72.3% | |
| All International Competitions | 292 | 57.8% | *p*-value: 0.0139 |
| All National Competitions | 709 | 61.0% | *p*-value: <0.0001 |
| Total | 1001 | 60.6% | *p*-value: <0.0001 |

**Table 5.2.** Determinants of Winner of Penalty Shoot-Out

| | Probit | Probit | Logit | Logit |
|---|---|---|---|---|
| Constant | -0.267 | -0.273 | -0.437 | -0.403 |
| | (0.217) | (0.506) | (0.343) | (0.609) |
| Team kicks first | 0.657*** | 0.633*** | 1.027*** | 1.012*** |
| | (0.140) | (0.134) | (0.192) | (0.187) |
| Home field | -0.092 | -0.114 | -0.128 | -0.165 |
| | (0.210) | (0.244) | (0.352) | (0.340) |
| Neutral field | -0.052 | -0.048 | -0.073 | -0.079 |
| | (0.275) | (0.314) | (0.422) | (0.412) |
| Category | 0.002 | -0.007 | 0.011 | -0.007 |
| (1 if higher) | (0.182) | (0.170) | (0.272) | (0.228) |
| "Team kicks first" interacted with | | | | |
| Home field | No | Yes | No | Yes |
| Neutral field | No | Yes | No | Yes |
| Category | No | Yes | No | Yes |
| *N* (teams) | 2002 | 2002 | 2002 | 2002 |
| Adjusted $R^2$ | 0.106 | 0.108 | 0.106 | 0.108 |

*Note*: Regressions in columns 2 and 4 also include fixed effects for Champions League, UEFA Cup, National Team, and National Cup competitions, as well as interactions between Home and Neutral field and Category.

the state of the competition itself and contribute to determining human performance in a strictly competitive (zero-sum) setting. What is not possible to identify, however, is the precise psychological mechanism that generates the result. We may speculate that the randomly determined order could generate differences in arousal, in anxiety, in shifting of mental process from "automatic" to "controlled," or in the narrowing of attention. Maybe it also generates differences in reference points. Köszegi and Rabin (2006), for instance, develop a model where a person's reference point is her or his rational expectation of the outcomes and "gain-loss" utility evaluations around this point influence her or his behavior. In a penalty shoot-out, the score at the time a player has to perform his or her task (the "ahead-behind" asymmetry caused by the order) may perhaps act as a reference point that has an effect on behavior.

Although we cannot answer the question of what is the specific psychological mechanism at play in this effect in performance, we can attempt to answer various related questions:

1. Are subjects aware of the advantage of going first?
2. Do they rationally respond to this advantage by systematically choosing to kick first when given the choice (after 2003)?
3. Do players talk about a specific psychological mechanism that is at work in generating these effects?

According to a survey conducted in AP (2010), the answer to the first two questions is affirmative (see table 5.3).

Clearly, if subjects are aware of the effect, they should always choose to go first. Unfortunately, there are no public records of players' choices because FIFA regulations do not require referees to record this information. By watching matches that end in a penalty shoot-out, it is sometimes possible (when the TV channel is not airing commercials), to catch the instant when the referee flips the coin and talks to the winner of the toss. Consistent with their answers in the survey, in every case when it was possible to see the coin toss, the winner of the toss was observed to choose to kick first, with just two exceptions. The first exception is the Italy–Spain match in the quarter-finals of the European Championship in June 2008. Gianluigi Buffon, the goalkeeper from Italy, won the toss against Iker Casillas, the goalkeeper from Spain, and chose Spain to kick first. Interestingly enough, the second exception involves the

**Table 5.3.** Survey

The following questions were asked to soccer coaches and players:

Q1: "Assume you are playing a penalty shoot-out. You win the coin toss and have to choose whether to kick first or second. What would you choose: first; second; either one, I am indifferent; or, it depends?"

Q2: "Please explain your decision. Why would you do what you just said?"

| | Observations | First | Second | Indifferent | Depends |
|---|---|---|---|---|---|
| Coaches | | | | | |
| Professional | 21 | 90.5% | 0 | 0 | 9.5% |
| Amateur | 37 | 94.6% | 0 | 0 | 5.4% |
| Players | | | | | |
| Professional | 67 | 97.0% | 0 | 1.5% | 1.5% |
| Amateur | 117 | 96.5% | 0 | 2.5% | 1.0% |
| All | 242 | 95.9% | 0 | 1.6% | 2.5% |

*Notes:* Professional coaches and players come from the professional leagues in Spain (Primera División and 2A and 2B División). Amateur coaches and players come from División 3 and regional leagues in Spain. The four coaches who answered "It depends" further explained that they would let their players choose what they preferred to do.

same teams and the same players five years later. In the semifinal of the Confederations Cup in June 2013, Casillas won the toss this time and decided to return the favor: He chose Italy to kick first. Perhaps goalkeepers are, after all, different from other players. Or as the old saying goes, you do not have to be crazy to be a goalkeeper, but it helps.

Finally, with regard to the third question, most subjects argue that their choice is motivated by the desire to put pressure on the kicker of the opposing team. Coding their answers to this question in the survey, in 96% of the cases they explicitly mention that they intend to put psychological pressure on the second kicking team. This is consistent with the evidence reported in AP (2010) that kickers decrease their performance when lagging (as opposed to goalkeepers, who improve theirs when leading).

*

A main difficulty for identifying the specific psychological mechanism at play is that a penalty kick involves two people, not one, and so the effect could arise from one player, from the other, or from both. An idea then is to look at similar sports settings that involve analogous dynamic decision-making processes but involve just one individual, not two, and also to look at other competitive activities with two individuals.

Mertel (2011) looks at data on more than 220,000 free throws from four seasons of professional NBA basketball. Carefully controlling for reverse causality, serial correlation, and a number of potential factors, he finds that players are significantly *more* likely to hit their free throws when they are *ahead* on the scoreboard than when they are behind. The difference in the scoreboard stops being relevant once the outcome of the game is beyond doubt and players revert to their inherent ability-reflecting mean. These findings are important because a free throw is an individual *nonstrategic* task and the results are consistent with the evidence from penalty shoot-outs: A leading or lagging asymmetry in a dynamic competition causes differences in performance.

In a golf setting, Pope and Schweitzer (2011) analyze more than 2.5 million putts in tournaments of the PGA Tour using precise laser measurements. They find that even the best golfers—including Tiger Woods—show evidence of loss aversion (Kahneman and Tversky 1979): Professional golfers hit birdie putts less accurately than they hit otherwise similar par putts. Golf provides a natural setting to test for loss aversion because golfers are rewarded for the total number of strokes they take during a tournament, and yet each individual hole has a salient reference point, par. When hitting a birdie, a player is "leading" over the hole, whereas when hitting a par, the player is "lagging" and has a chance to "tie" the hole. As indicated already, Köszegi and Rabin (2006) model a person's reference point as her or his expectations about outcomes, and gain–loss utility evaluations around this point influence her or his behavior. In golf, par seems a natural reference point, and in a penalty shoot-out it is possible to conjecture that the score at the time a player kicks acts as a reference point. Consistent with this reference-point hypothesis, the accuracy gap between par and birdie putts diminishes for very difficult holes and the gap between par and bogey putts widens for very difficult holes. A difficulty in this golf setting, however, is that risk taking and performance cannot be measured separately.

Perhaps the cleanest evidence showing that an interim rank (a leading–lagging asymmetry) in a dynamic competition affects performance comes from weightlifting, which, like a free throw in basketball, is an individual, nonstrategic task. Genakos and Pagliero (2012) empirically study the effect of interim rank on performance using data on professionals competing in tournaments for large rewards. The fact that risk plays a role in this setting would appear to make the empirical identification difficult. However, the authors observe both the intended action (competitors announce the weight they want to lift) and the performance of each participant, and so they can measure risk taking and performance separately. They obtain two important findings. First, risk-taking exhibits an inverted-U relationship with interim rank. Revealing information on relative performance induces individuals trailing just behind the interim leaders to take greater risks. Second, and most relevant in the context of this chapter, competitors systematically *underperform* when ranked closer to the top, despite higher incentives to perform well. In other words, disclosing information on relative ranking *hinders* performance.

Although the identification of the exact channel through which emotions affect performance remains an open question, these different results from other sports on nonstrategic tasks are consistent with the hypothesis that information on relative performance hampers performance by increasing psychological pressure when subjects are lagging in the competition.

The implications of this phenomenon may be wide ranging and perhaps extend to other areas. For instance, Heckman (2008) remarks that emotional skills help determine a number of socioeconomic outcomes, contribute to performance at large, and even help to determine *cognitive* achievement. Understanding whether or not psychological elements that determine performance in noncognitive tasks (kicking a soccer ball, weightlifting, golf, basketball) may also contribute to explaining cognitive performance is a fascinating issue. Needless to say, it would be ideal to study an identical setting (a sequential tournament competition between two people who play a two-person game with a randomly determined order) performing a cognitive task rather than a noncognitive task. Luckily, this setting exists.

In a chess match, two players play an even number of chess games, typically either eight or ten games, against each other. One game is generally played each day, with one or two rest days during the duration of the match. The basic procedure establishes that the two players alternate the colors of the pieces with which they play. In the first game, one player plays with the white pieces and the other with the black pieces. In the second game, the colors are reversed, and so on. Who begins with what color is randomly determined, and this is the only procedural difference between the two players. According to the rules of FIDE (the Fédération Internationale des Echecs, the world governing body of chess), the order is decided randomly under the supervision of a referee. This random draw of colors, which is typically conducted publicly during the opening ceremony of the match, requires that the player who wins the draw plays the first game with the white pieces, which are strategically advantageous.

Hence, as in a penalty shoot-out, an explicit randomization method determines which player begins playing in a given role in a sequence of tasks or games where both players have exactly the same opportunities to play the same number of times in the same role, have the same stakes, and where all other circumstances are identical. As a result, as in a shootout, we should expect that two identical players have exactly the same probability of winning the match. That is, there is no rational reason why observed winning frequencies should be different from 50–50 in a large sample of chess matches. Yet Gonzalez-Díaz and Palacios-Huerta (2012) find that this is not the case. Instead, winning probabilities are about 60–40 in favor of the player who plays with the white pieces in the first and all the odd games of the match, and hence is more likely to be leading during the match.

*

The empirical evidence in this chapter shows that information on the performance of competing agents during the competition has an effect on noncognitive (soccer, basketball, weightlifting) and cognitive (chess) performance. Thus, as competitive situations that involve performing both cognitive and noncognitive tasks are ubiquitous in real life, the results may have broad applicability. There are, of course, numerous *strategic* reasons why in a sequential competition the order may give advantage to either a *first* mover or a *second* mover (see, e.g., Dixit and Pindyck, 1994; Cabral 2002, 2003). What the results in this chapter show is that there are, in addition, *psychological* reasons why leading or lagging may affect the performance of the competing agents.

An important consequence of these results is that a randomly determined order, which in a sequential tournament competition is obviously fair from an ex ante perspective, need not be ex post fair if it gives any type of advantage to a subset of the competitors.

The question then is this: How should the order of a sequential tournament competition between two agents be determined to make it both ex ante and ex post fair? Is it possible to improve upon the perfectly alternating order? A simple idea would be to change the type of tournament: Instead of sequential, make it simultaneous (e.g, the two teams in a penalty shoot-out may kick simultaneously in the two goals of the field). A similar alternative would be to keep it sequential but provide no information about the state of the competition until all the competitors have performed the same number of tasks. After all, it is *knowing* that one is leading or lagging that affects performance. For obvious reasons, either one of these alternatives is typically unfeasible or unattractive in sports, auctions, and other settings.

So, what can be done? Consider a sequential tournament where two players or teams *A* and *B* play against each other an even number of times. Say that a fair coin selects *A* to perform his or her task first and *B* second in the first two rounds. What should the order in the next two rounds be to attempt to make it ex post fair? Is there a way to improve the ex post fairness of the strict alternation of the order of play *A B A B A B A B* . . .? Well, if the order *A B* offers *any* kind of advantage to *either* player, then by reversing the order in the next two rounds, we tend to compensate that advantage. Doing so means that the resulting sequence in the first four rounds is *A B B A*. And, of course, this reversing is innocuous if no advantage existed in the first place. How about the next four rounds? The same principle applies: By reversing the order followed up to that point, we tend to compensate any potential advantage that might have been given to either one of the players until then. The resulting sequence is *A B B A B A A B*. And, again, reversing the order is innocuous if no advantage existed in the first place, that is, if *A B B A* in the first four rounds already provides no ex post advantage to either player. Logically, we can apply the same principle ad infinitum and keep reversing the order followed from the beginning up to that point:

$$ABBABAABBAABABBA\ldots$$

This sequence is interesting, and it has a name: the Prouhet–Thue–Morse (PTM) sequence. Mathematician Axel Thue discovered it in Thue (1912) while studying avoidable patterns in binary sequences of symbols, e.g., 0 and 1. It is defined by forming the bitwise negation of the beginning:

$$\tau = 0110100110010110\ldots$$

where 1 is the bitwise negation of 0, 1 0 is the bitwise negation of 0 1, 1 0 0 1 is the bitwise negation of 0 1 1 0, and so on. Formally, the PTM sequence $\tau = (t_n)_{n \geq 0}$ is defined recursively by $t_0 = 0$ and $t_{2n} = t_n$, $t_{2n+1} = t'_n$ for all $n \geq 0$, where for $u \in \{0,1\}$ we define $u' = 1 - u$.

This sequence $\tau$ was already implicit in Eugène Prouhet (1851) and was later rediscovered by Marston Morse (1921) in connection with differential geometry. Worldwide interest in this sequence has developed during the past century as research has shown that it is ubiquitous in the scientific literature. In fact, this sequence occurs as the "natural" answer to various apparently unrelated questions, for instance, in combinatorics, in differential geometry, in number theory (e.g., the Prouhet–Tarry–Escott problem), in group theory (e.g., the Burnside problem), in real analysis (e.g., the Knopp function), in the physics literature on controlled disorder and quasicrystals, in music, in chess, in fractals and turtle graphics (e.g., the Koch snowflake), and in many other settings (see Allouche and Shallit (1999) for a survey).

Hence, the PTM sequence can also be the answer to an important problem in economics: How should the order of a sequential tournament competition between two agents be determined to make it both ex ante and ex post fair?

Unfortunately, the PTM ordering is not followed in tournament competitions, including major sports competitions, sequential auctions, and others. The closest we find is serving in tie-breaks in tennis where the order of serves one and two (*A B*) is reversed for serves three and four (*A B B A*), and then this sequence is repeated *A B B A A B B A A B B A* . . . until a player wins by a certain margin. The serving order in tennis would be perfectly fair ex post if any advantage given by the order in the first two serves, *A B*, is exactly compensated by having the order in the third and fourth serve reversed, *B A*. Of course, it is not known if this condition is empirically satisfied.

The PTM sequence, therefore, offers potential for improving the fairness of sequential tournament competitions.[9] It is important that the sequence has $2^{n+1}$ elements, $n \geq 0$, that is, that its first half is the negation of the second half. Otherwise, the full potential is not realized (e.g., in a soccer penalty shoot-out, the winner should be the best of $2^3 = 8$ penalty kicks or best of $2^4 = 16$, etc., not the best of 10 penalty kicks, as it currently is). Clearly, the margin of victory chosen to determine the winner is irrelevant for the ex post fairness of a sequence with $2^{n+1}$ elements.

9 Let $\Delta(\tau,n)$ denote the ex post difference in performance between the two identical subjects in a PTM sequence of $2^{n+1}$ elements, $n \geq 0$. Reversing tends to compensate any advantage if $|\Delta(\tau,n)|$ decreases with $n$. A necessary and sufficient condition for the PTM sequence to be ex post fair is that $\lim_{n\to\infty} \Delta(\tau,n) = 0$.

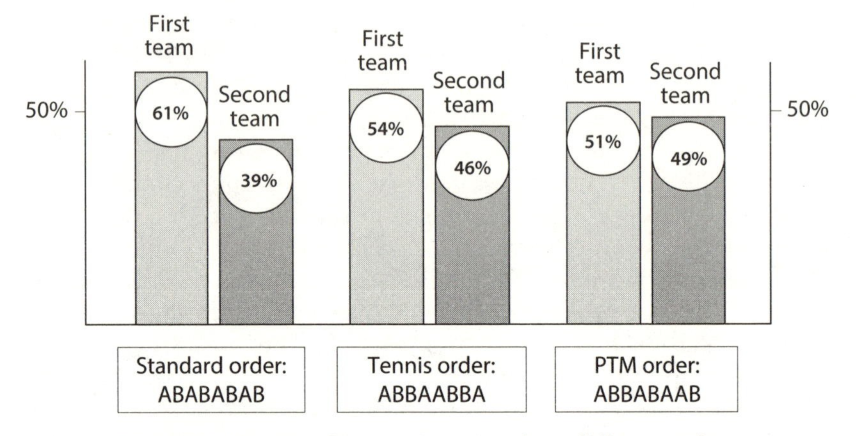


**Figure 5.2.** Winning frequencies using three different orders.

Since we are studying penalty shoot-outs in this chapter, it would be interesting to quantify the speed of convergence in this setting to produce an approximately fair outcome. How many rounds would be necessary to get "close enough" to 50–50? Is the number of rounds reasonable?

We explore this question with three experiments with professional players from Spain's La Liga (see figure 5.2). We implement penalty shoot-outs with three different kicking orders: In the first experiment, the sequence is *A B A B A B A B*; in the second, the order is the one followed in tennis: *A B B A A B B A*; and in the third, we follow the PTM sequence: *A B B A B A A B*. There are 200 shoot-outs in each experiment, each one involving 8 penalties, 4 per team, so that they can be perfectly compared. The order in each experiment is of course randomized.

The standard perfectly alternating order in the first experiment produces basically the expected advantage for the first kicking team: 61–39. Interestingly, when teams follow the tennis sequence, the advantage for the first kicking team decreases to 54–46, that is, from 22 percentage points it drops to just 8 percentage points. The advantage is further reduced if the PTM order is followed to just 2 percentage points: 51–49. Judging from these experiments, it appears that we do not need an excessive number of rounds to get reasonably close to 50–50, and so it seems quite feasible to improve the unfairness of the current perfectly alternating system in world soccer.

*

Sports competitions form an important class of fair division problems because sequences of strict alternation often give an unfair advantage to one competitor. This chapter has shown that the advantage may be not only substantial but entirely psychological. There are other problems of fair division that also have the same structure and have already invoked the PTM sequence. Brams and Taylor (1999) invoked this sequence, but did not identify it as such, when allocating a contested pile of items between two parties who agree on the items' relative values. They suggest a method called *balanced alternation*, or *taking turns taking turns taking turns*, as a way to circumvent the favoritism inherent when one party chooses before the other. Levine and Stange (2012) proposed the PTM sequence as a way to reduce the advantage of moving first when sharing a meal (more precisely, in the Ethiopian Dinner game, in which two players take turns eating morsels from a common plate). Richman (2001) had already studied such equitable resource allocation problems, but he too did not identify the sequence as such at the time of publication. More recently, Cooper and Dutle (2013) show that two duelers with identical lousy skills (known as "Galois duelers" in honor of the famous mathematician Évariste Galois, who was killed in a duel at the age of 20) will choose to take turns firing according to the PTM if they greedily demand their chances to fire as soon as the other's a priori probability of winning exceeds their own.

# HALFTIME

# 6

# SCORING AT HALFTIME

© TAWNG/STOCKFRESH

Michael Parkinson: "What was the nearest to kick-off that you made love to a woman?"

George Best: "Er—I think it was halftime actually."

George Best (May 22, 1946–November 25, 2005) was the first celebrity soccer player and "unquestionably the greatest British player ever" (Best 2004). In a captivating autobiography of the same title as this chapter, written in a breezy, self-deprecating style, Best (2004) gives us not only interesting suggestions for potential activities at halftime, but also ideas for how to spend time before and after games, too.

Nowhere in his autobiography, however, does Best appear to show an interest in the stock market. And why would he? He may well have been uninterested in investing. After all, why postpone current pleasures for the sake of potentially greater ones in the future? Perhaps he had even figured out that unless he had extraordinary insight or inside information, no stock would be a better buy than any other, and so there was little point in investing.

These are no more than conjectures, of course. During Best's career in the 1960s and 1970s, stock markets and betting markets were nowhere near as developed as they are today. It was difficult, perhaps even impossible, to buy and sell assets at halftime, or even to bet during soccer games.

This chapter is concerned with the idea of scoring at halftime but with a more scientific perspective. It turns out that what happens at halftime in some soccer games scores big in terms of allowing us to test a most influential theory in economics.

"Nothing in the toolbox of economists makes us good stock pickers. Yet we economists have written countless studies about the stock market" (Mankiw 2013). The noneconomist may be surprised to read a statement like this. But the most prominent theory of the stock market—the efficient-markets hypothesis—posits exactly that: The market processes information so completely and quickly that any relevant news would be incorporated fully into the stock's price before anyone had the chance to act on it. Simply put, unless one knew information that others did not know, no stock should be a better buy than any other (something which, incidentally, implies that expert money managers are not worth their cost).

The efficient-markets hypothesis is most commonly associated with Eugene Fama (1965, 1970, 1998), and its early origins can be traced back to Louis Bachelier (1900), who studied the dynamics of stock price behavior. If the theory is correct—that is, if observed changes in stock prices are unpredictable—there is not much we can do to gain an advantage over other traders, except perhaps to try to identify the news that causes stock prices to rise and fall and to understand the size of any likely price jump. Even this identification is difficult, often impossible.

If the efficient-markets theory is correct, then the price of an asset should jump up or down discretely when news breaks and then remain flat until further news arrives. To actually know if the theory is correct, we would need to isolate a news event—finding a meaningful window of time after it in which we could be certain that no further news has arrived. But how can we ascertain that no more news has occurred when there is the potential for news to arrive continually? It does not appear to be possible to stop the time for news but let the time for trading continue (and then test that the price is not changing). Time is the same for everyone.

Certainly finding an interval of time in which news cannot arrive but trading continues appears to be impossible in today's markets with modern communication technologies. Things were different in the past, though: If we look back far enough, we see that the time for news could be stopped, so to speak. Recent research by Koudijs (2013) exploits a beautiful historical case. The idea again is to study volatility in the absence of news to measure market efficiency. More precisely, if the underlying fundamentals of a stock traded in country *B* are known to happen exclusively in country *A* but somehow news travels with a lag from *A* to *B* (say, because of exogenous constraints on communication technology), then the price of the stock should not change until the news actually has arrived. If on the other hand, stock prices fluctuate more than a rational valuation of the underlying fundamentals would imply, then they should have a life of their own even in the absence of news.

During the 18th century, a number of English securities were traded on the Amsterdam exchange. Dutch holdings of English securities during the 1770s and 1780s represented between 20% and 30% of the total (Bowen 1989; Wright 1999), and so it was an important market. For the specific periods that Koudijs studies (1771–77 and 1783–87), virtually all relevant information originated in London and reached the continent via mail boats. In particular, there was an official packet boat service between London and Amsterdam that sailed twice a week. Importantly, news flows from London to Amsterdam were sometimes interrupted for exogenous reasons because bad weather could delay boats for days in a row. When no mail boats arrived, virtually no other relevant information reached the Amsterdam market. Using the exogenous breaks in the arrival of information and measuring price volatility during periods with and without news, Koudijs finds that security prices moved significantly even in the absence of news. In particular, around 20% to 50% of the overall return variance is unexplained by information. This result suggests that the Amsterdam market moved more than can be explained by the arrival of news, although the majority of price movements were still the result of efficient price discovery.

But again assessing market efficiency is not straightforward. Even in this clean historical case, the word "virtually" had to be used twice concerning the origin and the arrival of relevant information. As beautiful as this case is, it is still not possible to ascertain with certitude that all news relevant to the market originated in England and that absolutely no other relevant news from other sources arrived in Amsterdam. There are, in addition, other difficulties, such as those typically associated with historical data on prices, news, market structure, market participants, and trades. Data from historical markets are far less clean than data from today's markets. Also, of course, one may think that the main research interest concerns modern financial markets, which are an order of magnitude different from markets that existed more than 200 years ago.

The prospects for carrying out a similar investigation in modern markets, though, would appear to be hopeless. Information arrives continuously in today's markets, and so it appears to be impossible to study how much markets would have moved in the absence of news. Relating asset price fluctuations to the intensity of news arrival is also difficult because the latter is hard to measure. It is often unclear when specific news is observed and whether the information is relevant. In addition, a large fraction of information might be private, arriving in the market in the form of informed trades. We often do not know that relevant information has arrived until the market actually moves.

Perhaps unsurprisingly, and consistent with the motivation for this book, soccer provides a unique setting where none of these difficulties exist. A new article by Karen Croxson and James Reade (2013) provides an unusually clean study of market efficiency using high-frequency data extracted from live and heavily traded soccer-betting markets.

Live sports-betting markets also offer important advantages over most financial markets. Contracts on sports outcomes (unlike equities and other financial securities) have well-defined terminal values and converge to these values over a short period of time. Moreover, major sports news often breaks remarkably cleanly, becoming common knowledge at a single identifiable point in time. This phenomenon is particularly so where sports events are televised live, as these days a great many are.

The major news in soccer betting concerns the arrival of a goal. If betting markets are efficient, then prices should respond *immediately* and *completely* to goal arrival. However, assessing market efficiency in this setting is still complicated because even efficient prices would be expected to drift continuously during the game as traders update the passage of playing time, itself a stream of minor news. It follows that any observed postgoal price drift during the game would be difficult to interpret; it could reflect sluggishness in updating to major news (and hence evidence of inefficiency), or it could reflect an entirely efficient response to the ticking away of playing time.

This identification challenge is cleverly addressed by exploiting the existence of the halftime interval. In fact, this break in play provides a golden opportunity to study market efficiency. The reason is that the playing clock stops but the betting clock continues. This fact means that any drift in halftime prices can be interpreted unambiguously as evidence for market inefficiency since efficient prices should *not* drift during the news-free interval. This is a testable hypothesis. We can apply a test for statistical efficiency to halftime prices in games in which a goal arrives just before the start of the break, henceforth "cusp goals," as well as a test for economic efficiency, which would ask whether a hypothetical trader could make money during the interval by exploiting any potential over- or underreaction to cusp goals.

The key strengths of the halftime identification are simplicity and cleanness. A potential worry could be that perhaps trading is much quieter when the teams are in their dressing rooms. If there are no trades, then the price does not move and hence there is no volatility. Such a situation would not provide a test of market efficiency. Reassuringly, though, betting activity turns out to be healthy during the interval, even more so than during the game, and not least in matches with cusp goals.

The most common form of bet is a "fixed-odds" bet. Suppose that two bettors wish to take opposing sides to a bet—one wishes to back (bet on) an outcome, and the other wishes to lay (bet against) the same. Under a fixed-odds bet, the layer agrees to pay the backer a fixed multiple of the stake if the outcome takes place and gets to keep the stake otherwise. For example, George might feel very confident that Manchester City will win the F.A. Cup final against Wigan. He might offer Anna the chance to stake $10 at odds of 7:1 ("seven to one") that Wigan will win. In this case, Anna collects $7 \times 10 = \$70$ from George if Wigan succeeds, but otherwise George keeps Anna's $10 stake. Odds relate inversely to the probabilities associated with particular outcomes.[1] For instance, odds of 19:1 on the Dallas Cowboys to win the Superbowl would suggest that the market believes that the Cowboys are 19 times as likely to fail as to succeed; that is, they have a 5% chance of winning. In this example, the odds are quoted in so-called fractional form. An alternative is to quote decimal odds, in which case the stake is included in the quoted multiple, meaning fractional odds of 4:1 become decimal odds of 5.[2]

Until relatively recently, the betting markets were dominated by bookmakers, a closed community of licensed dealers. Akin to market makers in financial markets, bookmakers establish and maintain liquid markets in popular betting events by quoting prices at which they will deal. A bookmaker's customers are restricted to backing outcomes only; the bookmaker takes the lay side on every bet. Until recently, customers were also prevented from placing bets after the start of a sports event. A little over a decade ago, online betting exchanges began to emerge, mirroring the development of electronic exchanges for trading financial markets. Just as in financial markets, the emergence of exchanges proved to be an enormously disruptive innovation, transforming the betting experience for customers. The leading exchanges are essentially order-driven markets in fixed-odds bets. They allow individual customers to bet with each other directly, thereby disintermediating the

1 There is some debate about the interpretation of betting prices as probabilities. The interested reader is referred to Manski (2006),Wolfers and Zitzewitz (2006), and the articles cited therein.

2 Quoting odds in decimal form is convenient because the implied probability is then obtained simply by inverting the decimal odds and normalizing.

bookmaker. This situation means that exchange bettors are no longer restricted to backing outcomes; they can also lay them if they wish. In addition, exchanges began to offer customers the opportunity to place bets "in running," that is, once an event is underway. Betting during live events has become extremely popular. Typically, exchanges charge customers a small commission but do not impose a spread. Their prices have tended to be highly competitive, at least for popular events.[3]

The selection of markets offered by the dominant betting exchange Betfair is vast and covers an extensive range of sporting events. It also offers betting on political elections, reality TV outcomes, and other events of popular interest. Soccer recently surpassed horse racing as the biggest source of Betfair revenues. Within soccer betting, customers can place bets related to a range of outcomes, including the outright winner of a particular league or tournament or the top scorer of the competition. "Match Odds" markets allow betting on the outcome of individual games, by backing (betting on) or laying (betting against) the home win, away win, or draw.

Figure 6.1 shows the Betfair order book for a past Premiership soccer match between Arsenal and Manchester United. This snapshot was taken shortly before kickoff.

Suppose the user wishes to back Manchester United to win. According to the order book, the user might immediately stake up to $36,784 at odds of 3.3, and up to a further $53,140 at slightly less attractive odds of 3.25. Betfair uses decimal odds, which are inclusive of stake. So a $10 bet to back Arsenal at odds of 2.58 would result in a gross return of $25.80 ($15.80 profit plus $10 stake). All odds are displayed from the backer's point of view. Thus, 2.6 and $36,289 on the lay side of that market implies that someone (or some combination of users) has submitted limit orders hoping to back Arsenal at odds of 2.6 (i.e., slightly better than the prevailing market odds). If she or he were to accept $10 of this "volume" by placing a lay order at 2.6, the user would be betting against an Arsenal win and risking $26 to win $10.

Croxson and Reade use a data set comprising second-by-second snapshots of Betfair's live order book for 1,206 professional soccer matches in running, that is, as the match is being played. The sample of games spans a wide range of competitions: domestic, international, club, and national team matches. The order book shows prices and volumes from Betfair's Match Odds markets. Match Odds markets for professional

3 Ozgit (2005) finds that the exchange's basketball prices are more attractive than those of bookmakers but that its markets sometimes fail to offer deep liquidity at inside (best) prices. Croxson and Reade (2011) find that the largest betting exchange Betfair offers the best prices for betting on soccer up to relatively large bet sizes (over $800).

Betfair Soccer >> Arsenal v Man Utd

| Arsenal v Man Utd | + | | | | | Options ➤ |
|---|---|---|---|---|---|---|
| Change: Express view \| Full view | | | Matched: USD 4,511,471 | | | Refresh |
| Selections: (3) 100.3% | | | Back | Lay | | 99.1% |
| Arsenal | 2.54<br>$22976 | 2.56<br>$30417 | 2.58<br>$132 | 2.6<br>$36289 | 2.62<br>$17123 | 2.64<br>$2156 |
| Man Utd | 3.2<br>$39797 | 3.25<br>$53140 | 3.3<br>$16784 | 3.35<br>$40198 | 3.4<br>$14483 | 3.45<br>$7268 |
| The Draw | 3.1<br>$10541 | 3.15<br>$22637 | 3.2<br>$73984 | 3.25<br>$44782 | 3.3<br>$83524 | 3.35<br>$61519 |

**Figure 6.1.** Betfair order book for Arsenal vs. Manchester United.

soccer matches tend to be heavily traded, particularly when the matches are in progress ("live betting"). Across the sample as a whole, the average match sees more than $6 million staked on the three basic outcomes of the game. Typically, half of this volume is bet in running, which equates to $31,627 traded per minute and $527 per second. Behind this headline average, the betting interest is quite variable, with betting volume of $50 million in the most heavily traded match, compared with a little more than $0.05 million in the least traded. Many English Premiership games are televised, and the sample features an interesting mix of televised and untelevised encounters. Television coverage tends to be associated with significantly higher trading volumes.

An important and, for our purposes, useful characteristic of Betfair is that it briefly suspends its in-play soccer markets at kickoff and then briefly again upon the arrival of a "material event," such as the scoring of a goal, the award of a penalty, or the dismissal of a player. As far as Match Odds markets are concerned, the scoring of a goal is the most important piece of news. Goals arrive fairly infrequently; there are on average 2.55 goals per match in the sample. During a goal-related trading suspension, Betfair discards any unfilled orders, thereby clearing out the entire betting order book. When the order book reopens, the odds have shifted, reflecting updating by the market about the relative chances of the home win, away win, and draw.

If Betfair markets are efficient, then prices (and the probabilities that these prices imply) should update to public news rapidly and fully. It is straightforward to check whether or not prices respond immediately to the news of a goal. Looking across the whole sample of about 2,500 goals,

Croxson and Reade find that the scoring team's probability of winning jumps up almost immediately by an average of 22 percentage points. The exact size of the shift varies markedly across games, depending on such factors as the current score, how long is left to play, and whether the scoring side is the favorite, underdog, and home or away team. Game-changing goals that occur toward the end of the game tend to have the greatest effect, as we would expect. For example, a goal scored after the 80th minute that gives the scoring side the lead on average boosts their implied probability of winning by 64 points.

But do these jumps in price represent complete updating to goals or simply the beginning of an inefficient adjustment process? As noted already, some drift in prices is perfectly consistent with, and indeed evidence for, market efficiency since efficient prices would be expected continually to update to the passage of playing time. This critical identification challenge is tackled by exploiting the news-free window of the halftime interval. In games where goals arrive on the cusp of halftime, prices should be flat during the subsequent 15-minute interval. Because time-related drift cannot be present during this period, the halftime break provides a unique opportunity to test cleanly for news-related drift.

The data set contains 160 cusp goals that arrive within five minutes of the end of the first half. Figure 6.2 shows the distribution of such goals in the sample. Home goals (*H*) account for 76 of the cusp goals; the other 84, therefore, are away goals (*A*). Favorites (Fav) score 103, and outsiders (Out) the remaining 57. A good number of goals are scored extremely close to the end of the first half; 53 arrive in the final minute of

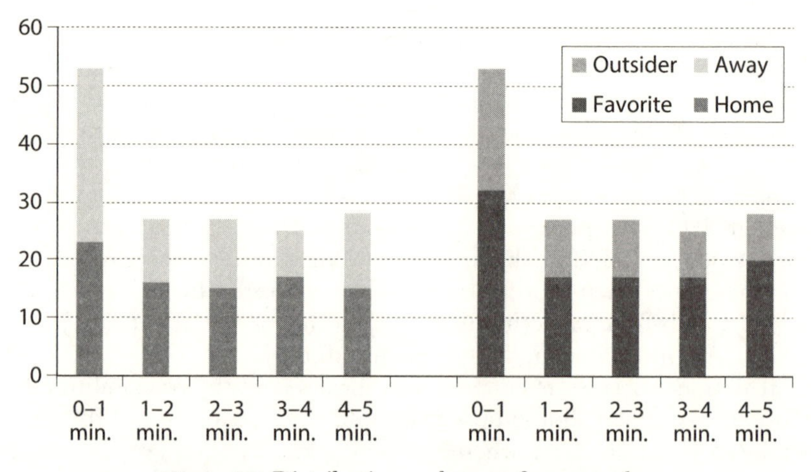


**Figure 6.2.** Distribution and type of cusp goals.

first-half play; a further 27 arrive in the penultimate minute. This relative abundance of goals on the edge of the break is very helpful: The closer the goal is to halftime, the stronger the efficiency test.

Before we get into the statistical and economic tests of market efficiency, let us do a visual inspection of the price data for a few of these matches. This inspection may give us a sense of whether we could expect the tests to provide evidence of efficient or inefficient updating.

Consider figure 6.3, which shows betting data for a match between Tottenham Hotspur (the home team) and Manchester United. The left-hand panel plots the probability of a Manchester win, as implied by the best Betfair back price. At the start of the match, this probability is 56% (Manchester is the favorite to win), but we observe this probability drift downward as the first half progresses without a goal. By the 44th minute, it has fallen to under 50%. Then, just before halftime begins, Manchester scores to go ahead, and the market is suspended briefly. When it reopens moments later, the new probability for a Manchester victory is 77%. Over the subsequent 15-minute break in play, does the implied probability appear to remain constant at this new 77% level? This visual inspection suggests that it does, and remarkably so. The updating to the goal appears to be immediate and complete. Also, the right-hand panel confirms that the market is traded actively throughout halftime; indeed, trading interest appears to increase somewhat during the halftime interval.

In figure 6.4, a cusp goal causes an upset; the prematch favorite goes down a goal just before halftime. Again there appears to be no obvious trending over the break, and here too trading is strong during the break in play. Trading is of course critical because we do not want to test for market efficiency when participants are not trading. Importantly, across all matches in the sample, the average volume traded per second is $527.

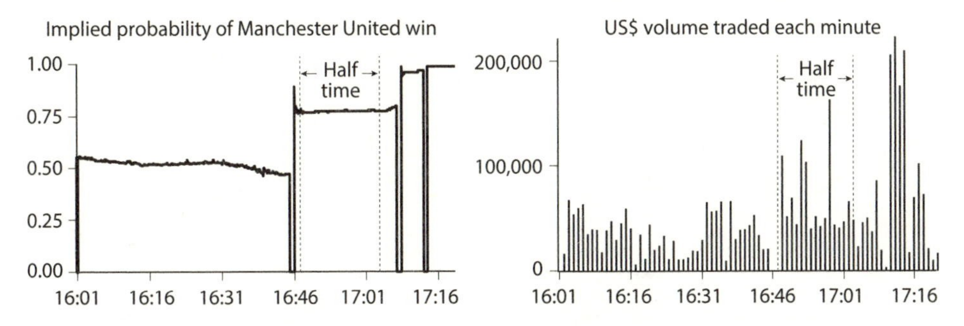


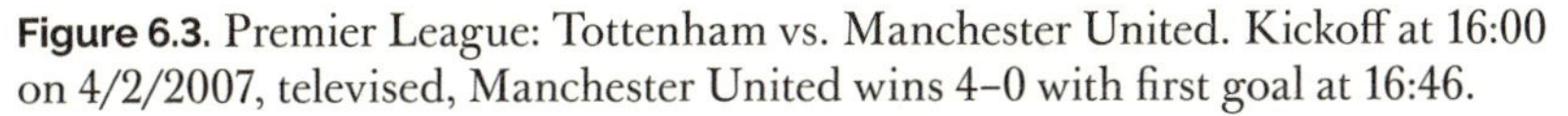
**Figure 6.3.** Premier League: Tottenham vs. Manchester United. Kickoff at 16:00 on 4/2/2007, televised, Manchester United wins 4–0 with first goal at 16:46.

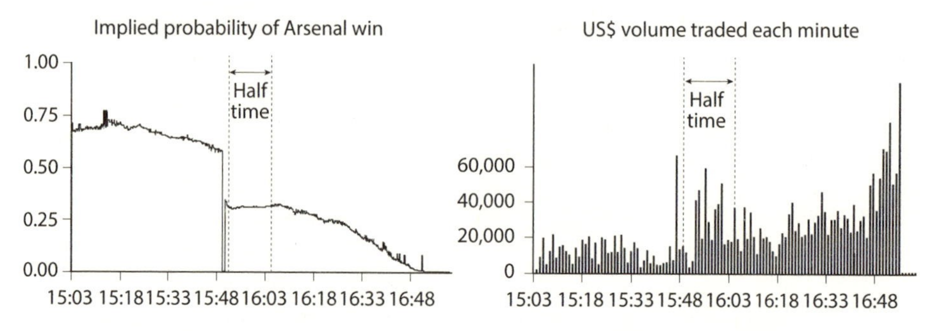


**Figure 6.4.** Premier League: Arsenal vs. West Ham. Kickoff at 15:00 on 7/4/2007, not televised, West Ham United wins 1–0 with goal at 15:50.

In games featuring a cusp goal, halftime trading volume averages $552 per second, a 5% increase.

Figures 6.5 and 6.6 provide two additional examples of prices and trading volume in matches featuring cusp goals, one from the Champions League (the famous final discussed in chapter 1 between Manchester United and Chelsea that was decided in a frantic penalty shoot-out) and one from Euro 2008, Spain vs. Russia. In both games, there were two goals before halftime (1–1 and 2–0). Again, there are very few signs, if any, of any trending in prices over halftime, and halftime trading volumes are healthy.

The apparent stability of halftime prices in games with cusp goals is indicative of market efficiency. To test this formally, Croxson and Reade implement two approaches. First, they use a regression analysis to identify the presence or otherwise of drift in halftime prices—this is a test for *statistical efficiency*. Second, they test for *economic efficiency* by exploring whether customers could make positive returns over the halftime interval by exploiting any systematic drift in prices.

The first step in testing for efficiency statistically is to construct an appropriate model of prices over the halftime break. If efficiency holds, the news of a goal scored in the first half should not have any predictive power when it comes to forecasting changes in prices over the halftime interval. All first-half news should be reflected in prices already, even where a goal arrived on the cusp of the break. The regression model should thus test whether during the halftime interval current prices can be forecast on the basis of lagged prices and check whether this lack of forecastability is altered by the arrival of a cusp goal.

Pooling the halftime price data across the full sample of matches, the following regression model can be estimated for each contract type $m$, where $m \in \{h,a,d\}$:

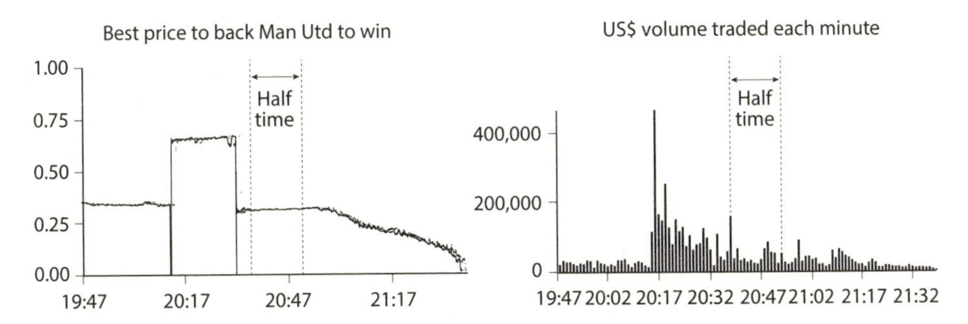


**Figure 6.5.** 2008 Champions League Final: Manchester United vs. Chelsea. Kickoff at 19:45 on 5/21/2008, televised, 1–1 draw after 90 minutes, first goal at 20:30:45.

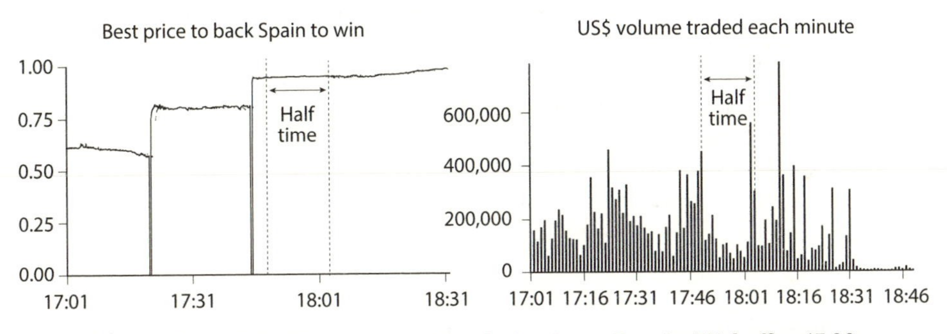


**Figure 6.6.** Euro 2008 Group Stage Match: Spain vs. Russia. Kickoff at 17:00 on 6/10/2008, televised, Spain wins 4–1 with cusp goal at 17:45:00.

$$pr_{c,m,t} = \theta_0 + \theta_1 pr_{c,m,t-1} + \theta_2 pr_{c,m,t-2} + \varepsilon_{c,m,t}$$

where $pr$ denotes the volume-weighted average price and the subscripts $m \in \{1, \ldots, M\}$ denotes the particular match, $c \in \{1, 2, 3\}$ denotes the contract traded (Home Win, Away Win, Draw), and $t \in \{1, \ldots, T\}$ denotes the second of time on the clock.

For each contract type, the authors have a panel data set composed of prices for the contract type in question over the $M = 1{,}206$ soccer matches. Under the null hypothesis of market efficiency, all coefficients in the above equation should be zero except for the first lag of price, $\theta_1$, which would be significant and take a value of unity:

$$\mathbf{H}_0 : \theta_0 = \theta_2 = 0, \theta_1 = 1$$

That is, all the information should already be included in the price the previous second, and no other variable should have any explanatory

**Table 6.1.** Critical Values for Rejection of the Null Hypothesis and Test Statistic

| Time Interval | Confidence Level | Im–Pesaran–Shin Tests Critical Values for Rejection | | |
|---|---|---|---|---|
| | | Home | Away | Draw |
| 5 minutes | 1% | 6.13 | 6.13 | 6.13 |
| | 5% | 4.51 | 4.51 | 4.51 |
| 10 minutes | 1% | 16.77 | 16.77 | 16.77 |
| | 5% | 3.73 | 3.73 | 3.73 |
| Test Statistic 5 min | | 6.44 | 6.92 | 6.78 |
| Test Statistic 10 min | | 9.59 | 10.68 | 9.78 |

power. Working with a generalized version of this model, which is then transformed for unit root testing, it is then possible to use the Im–Pesaran–Shin (IPS) test to evaluate the null hypothesis.

The critical values for rejection at the 1% and 5% levels when this test is applied to the first 5 minutes of all halftime prices are shown in the top half, and for the first 10 minutes in the bottom half of table 6.1. The actual test statistics appear in the final row of the table.

By comparing the test statistics with the rejection values, we see that it is not possible to reject the null hypothesis of market efficiency at the 5% level for the first 10 minutes of halftime, and at the 1% level for the first 5 minutes. Thus, consistent with the null hypothesis of market efficiency, the coefficient for the first lag of price is 1, and all other coefficients are zero at those confidence levels.[4]

The second and ultimately most meaningful test of market efficiency is whether customers could trade profitably on any potential over- or underreaction to news—in other words, whether the markets are economically efficient: Could customers make positive returns by exploiting any systematic drift in prices over the halftime interval?

Consider two hypothetical trading strategies:

1. Backing (buying) a particular match outcome at the start of the halftime interval and laying this (selling it back to the market) later in the break. This method would exploit any systematic downward

4 Other tests yield similar results. An interesting question is if these results can be generalized beyond the halftime interval to the game in play. Perhaps different types of traders are active during the halftime interval; perhaps major news interacts with more minor news during minutes of play. Croxson and Reade develop two robustness checks for testing the market's ability to update to the news of a goal while the match is in progress. These additional statistical tests also support the view that drift in Betfair prices *during* minutes of play is largely explained by efficient updating to the passage of playing time.

drift in odds during the break (perhaps caused by an initial under-reaction to a goal).
2. Laying (selling) a contract at the start of the halftime break and backing this contract at a later point during the interval. This strategy would exploit any systematic upward movement in odds during the interval (perhaps reflecting an initial overreaction).

The potential profitability of each strategy is investigated using a difference in means test. Let $p_{m,b,i}$ be the best back price for a particular outcome in minute $m$ of the halftime interval in match $i$, and let $p_{m,l,i}$ be the best lay price for the same outcome. Denote their respective means across all matches in the sample as $\varphi_{m,s}$, where $s \in (b,l)$. At any point in time, the best available price to back a particular outcome must lie below the best available price to lay the same outcome; otherwise, the exchange could immediately match some of the orders in the book by crossing trades at prices in between these. Consequently, it will be possible to profit from strategy 1 only if the best lay price after $X$ minutes of the interval has fallen below the initial back price: $p_{X,l} - p_{1,b} > 0$. A suitable difference in means test would calculate the $t$-statistic:

$$t = [\varphi_{1,b} - \varphi_{X,L}]/\sigma(\varphi_{1,b} - \varphi_{X,l})$$

where $\sigma(\varphi_{1,b} - \varphi_{X,l})$ is the standard deviation of the difference in means. Croxson and Reade implement this test for both strategies across the full sample considering both 5-minute and 10-minute trading intervals. In all cases, the results imply a clear rejection of the null hypothesis that the difference in means is positive. Neither strategy is profitable because the average decimal odds that must be offered in the lay trade are always strictly greater than those available for the back trade, implying a *negative* return.

Overall then, the halftime analysis yields conclusive evidence that Betfair markets are economically efficient: Prices impound news so rapidly and completely that it is not possible to profit from any potential price drift over the halftime interval.

*

The question of whether real-world markets are efficient has long engaged academics, policy makers, and practitioners alike. Efficient markets respond rapidly and completely to news, ensuring that the price of any asset reflects its true fundamental value at all times. Inefficient markets undermine the optimal allocation of resources, with adverse consequences for welfare; they also imply the potential for systematically profitable trading strategies. Perhaps unsurprisingly, extensive efforts have been made to ascertain the efficiency of real-world markets.

Most previous studies have examined financial markets—for instance, by analyzing the response of share prices to stock splits, the release of company results, merger announcements, and announcements about economic variables such as the money supply. The results overall have been mixed: Some studies have found support for market efficiency, whereas others have uncovered evidence that prices appear to drift after the arrival of news, indicative of an inefficient market.[5] As the press release for the recent Nobel Prize in Economic Sciences in 2013 awarded to Eugene F. Fama, Lars P. Hansen, and Robert J. Shiller indicates, our current understanding of asset prices relies in part on fluctuations in risk and risk attitudes and in part on behavioral biases and market frictions (official website of the Nobel Prize, http://www.nobelprize.org/nobel_prizes/economic-sciences/laureates/2013/press.pdf).

Unfortunately, studies of efficiency in financial markets typically suffer from a number of important limitations, including the fact that it is often difficult to ascertain when exactly news breaks and to know how much information is contained in the absence of news and the passage of time. The critical problem that even the absence of news and the passage of time often contain information is nicely illustrated in "Silver Blaze," one of the most popular Sherlock Holmes short stories written by British author Sir Arthur Conan Doyle (1892). This story focuses on the disappearance of a famous racehorse on the eve of an important race and on the apparent murder of its trainer:

> Gregory (Scotland Yard detective): Is there any other point to which you would wish to draw my attention?
> Holmes: To the curious incident of the dog in the night-time.
> Gregory: The dog did nothing in the night-time.
> Holmes: That was the curious incident.

There is also the difficulty of defining normal returns. Any test must assume an equilibrium model that defines normal security returns, but this assumption means that efficiency could be rejected because the market is inefficient or because the assumed equilibrium model is incorrect. This *joint hypothesis problem* means that market efficiency as such can never really be rejected.

A number of studies have examined informational efficiency in sports betting. However, live exchange-based betting is a recent phenomenon, and most previous analyses have relied on low-frequency bookmaker prices sampled before the start of a live event.

5 Vaughan Williams (2005) provides a comprehensive review of the academic literature on information efficiency in financial markets.

The question of whether betting markets are efficient has implications for the reliability of information markets (also called prediction markets). Information markets essentially are betting markets designed specifically to produce forecasts of future events (Wolfers and Zitzewitz 2004; Hahn and Tetlock 2006; Vaughan Williams 2011). Interest has grown over the past decade in the potential for such markets to improve decision-making across a wide range of settings. The novel idea in Croxson and Reade of using the break in play to separate event time from trading time is incredibly useful; it allows us to separate for the first time *efficient* game time-related drift from potentially *inefficient* drift in prices. The evidence shows that these markets are economically efficient.

Soccer is often called a game of two halves, but as George Best knew, interesting opportunities can also arrive at halftime.

# SECOND HALF

7

# FAVORITISM UNDER SOCIAL PRESSURE

© TAWNG/STOCKFRESH

Last time we got a penalty away from home,
Christ was still a carpenter.
—Robin M. Lawrence, Crystal Palace manager 2012

Social environments influence individual behavior. This important aspect has long been the focus of the literature on endogenous preference formation but only where convincing empirical tests are difficult to find. This chapter is concerned with the effect of nonmonetary incentives on behavior, in particular with the study of social pressure as a determinant of corruption. The analysis differs from extensive work in the literature on corruption both in the origin of the incentives to deviate from honest behavior (social pressure) and in the agent whose behavior is studied (a judge).

Few people receive as much pressure as some judges—in particular, the judges of soccer games. The pressure they experience ranges from the social to the divine, from the public to God.

The first source of pressure brings to mind the short story "Poor Dear Mother" by Eduardo Galeano (1995). In the late sixties, after a long absence from Ecuador, poet Jorge Enrique Adoum returned to his country. As soon as he arrived, he complied with the mandatory ritual in Quito: He went to the stadium to see his team Aucas play. It was an important game, and the stadium was packed.

Before the game began, there was a moment of silence for the referee's mother, who had passed away on the eve of the game. All the spectators silently stood up. Then, someone from the city hall delivered a speech highlighting the exemplary attitude of the referee who was about to do his duty in the most unfortunate circumstances. In the middle of the pitch, head down, the man in black received the warm applause of the audience. Adoum blinked; he pinched his arm. He could not believe what he was seeing. In what country was he? Many things had changed during his absence. It used to be that people only cared about the referee to yell at him: "Son of a bitch!"

And the game began. Fifteen minutes later, the stadium exploded in joy: Aucas scored! But the referee disallowed the goal because a player was offside. Immediately, the crowd reminded the referee who was the late author of his days: "Orphan of a bitch!" roared the stands.

On the pressure soccer referees get from God, Pedro Escartín (August 8, 1902–May 21, 1998) can provide solid testimony. He was a professional soccer player, coach, journalist, and from 1928 to 1948 also an international referee. He went on to become one of the most prestigious referees in Europe in the 1930s and 1940s. In 1940, he became a member of the FIFA Disciplinary Committee for 27 years, and in 1967 he received the FIFA Order of Merit. His last refereeing was a friendly match between Italy and England in 1948, the year in which he retired. Escartín was also a religious person, so knowing that he would be in Rome for a few days, he asked Pope Pius XII for an audience. Although it was something almost impossible to get, surprisingly, the audience was granted for the day after the match.

The final score was 4–0 for England, and Escartín disallowed two goals for Italy. The following day, he went to the Vatican. It was customary in papal audiences that people should be on their knees until the Pope himself were to allow them to stand up. When his turn arrived, Escartín obliged as everyone else and bent on his knees. Pius XII asked him, "And who are you?"

He replied, "Your Holiness, I am the referee of the match between England and Italy yesterday."

Slightly annoyed, Pope Pius XII protested: "But man, please, you are the one who disallowed two goals for the Italians."

To which Escartín replied, "Yes, your holiness, but they were justly disallowed." Apparently unconvinced, the Pope required Escartín to stay on his knees during all the audiences for the day. The Pope allowed him to stand up and leave only when everyone else was already gone, "as punishment of course (laughs). It was the biggest punishment I have ever received."[1]

1 See "Pío XII Sancionó a Pedro Escartín" in http://www.diarioninformacion/opinion/2010/07/12/pio-xii-sanciono-pedro-escartin/1026311.html, July 12, 2010, and

*

How important is social pressure as a determinant of decision-making? Can it corrupt the way we act? The intuitive answer would perhaps be affirmative, but until very recently there were no empirical studies on this important phenomenon.

There is a wide-ranging theoretical literature on the mechanisms of corruption, especially monetary ones like bribes or promotions, that features quite prominently in the development economics context. Likewise, the issue of corruption caused by inherent biases has also been explored on the economics of firms when supervisors have inherent preferences for certain workers. In recent decades, economists have begun to systematically explore the influences of social factors and environments on individual behavior. The desire for social approval and other interdependencies of human preferences with social forces have been used as explanations for a wide range of socioeconomic phenomena, from consumption behavior and cultural practices to parents' influence on children's preferences. For example, social norms supported by social approval have been posited as an explanation for equilibrium wages above the market clearing rate.[2]

For obvious reasons, however, it is difficult to test empirically theories that incorporate the effect of social influences on individual decision-making. Social forces are difficult to quantify or even to observe accurately, and the influence on behavior cannot be ascertained unless it is clear how the individual would have acted in the absence of such forces. These difficulties increase by an order of magnitude if we consider situations where the individual is interested in *hiding* his or her own behavior, such as corruption. It is no surprise that until recently, there simply was not even a single empirical study of the effects of social pressure on corruption. This is the topic of this chapter.

As in other chapters in this book, professional sports present a valuable setting for studying what might seem to be an impossible problem. Whereas testing for the effect of social pressure would normally require the researcher to disentangle complicated interacting effects, the environment of professional sports provides an ideal setting. In particular, a critical advantage is that in soccer and other sports most behavior is difficult to hide from observation.

---

the interview by Juan Adarve of Escartín's son "Pedro Escartín, la Leyenda que Cambió el Futbol," in http://www.guadanews.es/noticia/2221, March 24, 2011.

2 See Akerlof (1980), Bernheim (1994), Becker and Murphy (2000), and other references therein. DellaVigna (2009) surveys recent field studies on social preferences and group pressures that find evidence of their importance in such disparate areas as charitable giving, workplace relations, and fund-raising.

The setting concerns the effect of the preferences of a group (namely, the spectators attending a soccer game) on the behavior of the judge of the game. Professional soccer games are attended by huge crowds of up to 100,000 people in the top European and South American leagues, often overwhelmingly and loudly rooting for the home team. Does the referee "internalize" the social preferences in the stadium? Do these forces push him to rule in favor of one team over the other?

To anyone who follows soccer, the idea that the referee may be biased in favor of the host team sounds sensible. Casual observation suggests that for many types of decisions, from awarding penalty kicks to applying the offside rule, referees often seem to apply two different measuring rods. Yet it is often unclear, sometimes even impossible, to determine whether the referee's decision was just or unjust, correct or incorrect, even if it seems clear to the casual observer (especially if he or she is a supporter of one of the two teams). Many decisions require *subjective* judgment, and a referee must make decisions that would be considered dubious in retrospect even if the crowds were absent from the stadium.

Fortunately, there is one specific decision that a soccer referee makes that allows for a clean testing of the importance of social pressure in decision-making: the amount of *injury time* added at the end of the game.

Injury time is the amount of extra time added to the first and second halves of a soccer game to compensate for lost time caused by injuries, goals, and other unusual interruptions. Importantly, the official *Laws of the Game* (FIFA 2012) prescribe the reasons for such extra time. Because the outcome is easily quantified (extra time in minutes) and the decision should depend on events that are observable (the number of yellow and red cards, substitutions, etc.), it is possible to use this decision to test for the influence of social pressure.

Garicano et al. (2001, 2005) do precisely this. They take advantage of this setting to test for systematic bias using data from the Spanish professional soccer league La Liga. The results are quite stark: Referees on average add *more* injury time (controlling for a number of factors) when the home team is *behind* in a close game than when it is ahead in an equally close game. When the game is not close and extra injury time is less likely to change the outcome, no such bias is found. Moreover, using an exogenous change in the rewards for winning a game, they find that the higher the rewards, the greater the referees' bias. Finally, they also gather evidence on the channels through which social pressure influences the referees' decisions.

This chapter is concerned with these results. The logic proceeds in three steps. First, we show and quantify the referee's bias. The premise is that the amount of extra time should *not* systematically depend on the identity of the team that is leading at the end of a game. Yet it does, but only for close contests. On average, injury time is approximately 3 minutes. However, if the home team is behind by 1 goal, the injury time is 35% above average, whereas if it is ahead by 1 goal, the injury time is 29% below average. This difference only arises when the game is close: When either side is ahead by 2 goals or more, there is no significant difference.

Therefore, referees appear to use their discretionary power to favor home teams, but only in close games, when the time added has a reasonable chance to affect the final outcome. Moreover, although it might seem plausible that the game may be simply more intense when the away team is leading, controlling for factors that directly predict the intensity of the game (such as the number of disciplinary sanctions, player substitutions, or the strength of the teams involved in the contest) makes no difference to this result. This evidence gives strong primary evidence of bias.

Second, the hypothesis that referees show a bias for the home team because of social pressure means that the bias should be stronger when the crowd's rewards from winning are higher. This notion yields the testable prediction that when the home team has more to gain from a victory, the referee should become more biased. Accordingly, the second exercise is to show that referees show more favoritism when the returns to the crowd increase. To do so, the analysis exploits an exogenous change in the rewards for winning occurring during the sample period. Before 1995, a win was worth 2 points, a tie 1 point, and a loss 0 points. After 1995, the points awarded to winners increased from 2 to 3. As predicted, the evidence shows that after 1995, referees became more biased where the home team was ahead by 1 goal compared to when it was behind by 1 goal.

Third, what is the specific mechanism that could plausibly underlie this behavior? The hypothesis underlying this chapter is that it is the actual crowd *in the stadium* that puts pressure on referees. Although millions of people may care strongly about the outcome of the game, it is the (on average) 25,000 spectators around the pitch who are influential. The people at home are invisible and exercise no direct pressure. To test if this hypothesis is true, we can examine the connection between referee bias and the *size of the crowd*. What we find is that when crowds are larger, referees become more biased: An increase of one standard deviation in crowd size causes the home bias to rise by 20%.

But crowds do not always exclusively support the home team. Sometimes, a substantial part of the crowd supports the away team, also exerting social pressure on the referee. This variation in the support for the team can be used to investigate the relationship between the *composition of the crowd* and the amount of bias. Consistent with this intuition, when the crowd is likely made up of a substantial number of fans supporting the visiting team, the referee's bias in favor of home teams is mitigated. Thus referees react to the preferences of the representative supporter in the crowd.

The data come from one of the main professional soccer leagues in Europe (La Liga in Spain), where 20 teams play each other twice during the season, once as a home team and once as a visitor. A season lasts for approximately 9 months (September through May), and teams typically play one game per week. As is well known, at the end of each of the two 45-minute halves, the referee may award injury time to make up for the time lost during the game. Time awarded ranges in the sample from 0 to 7 minutes.

As in every league competition, the points teams receive determine the incentives they face. Three outcomes are possible: a win, a tie, and a loss. Until 1994–95, these three outcomes were respectively rewarded with 2, 1, and 0 points. After that season, the reward for winning was changed to 3 points. As noted already, we study the effects of this change in incentives on referee behavior using data from the 1994–95 season (380 games), the last one with the 2–1–0 reward scheme, and from the 1998–99 season (380 games) with the new 3–1–0 reward scheme. Studying four seasons apart is convenient because it does not require us to assume that teams were able to adjust immediately to the new situation.

Before we go on to test our hypotheses, note some descriptive statistics (see table 7.1). On average, there are 2.57 goals per game; the home team scores approximately half a goal more than the away team. Average crowd size is 28,000, but attendance can be as high as 98,000. Referees can discipline players for foul play in two easily observable ways: a yellow card, which allows the player to continue playing in the match unless he or she receives a second one, and a red card, which leads to immediate expulsion. On average, 4.78 yellow (2.5 of these to the away team), and 0.17 red cards are awarded per game.

Most interestingly, of course, is that referees add on average 2.93 minutes of injury time in the second half of the game and 0.79 minutes in the first half. The discretion that referees have over the amount of injury time varies in the sample. Until the World Cup of 1998, referees were free to add on as much extra time as they saw fit and had to notify nobody about the amount they intended to add on. Beginning in the 1998–99 season, however, the world governing body of professional soccer, FIFA, forced referees to publicly announce the intended amount of injury time at the end of normal play (in a way exacting a commitment).

Plotting average injury time played against the score margin at the end of the second half of play reveals some initial evidence of bias. For games with a difference of 2 or more goals in the score, the referee adds roughly the average amount of injury time, regardless of whether it is

**Table 7.1.** Descriptive Statistics

| Variable | Obs. | Mean | Std. Dev. | Min. | Max. |
|---|---|---|---|---|---|
| Score Difference | 750 | 0.58 | 1.71 | –5 | 6 |
| Score Home Team | 750 | 1.57 | 1.32 | 0 | 7 |
| Score Visiting Team | 750 | 1.00 | 1.08 | 0 | 7 |
| Goals in Extra Time, Home | 750 | 0.04 | 0.21 | 0 | 1 |
| Goals in Extra Time, Visitor | 750 | 0.03 | 0.17 | 0 | 1 |
| Minutes Extra Time, 2nd Half | 750 | 2.93 | 1.11 | 0 | 7 |
| Minutes Extra Time, 1st Half | 750 | 0.79 | 0.73 | 0 | 3 |
| Yellow Cards Home | 750 | 2.23 | 1.37 | 0 | 7 |
| Yellow Cards Visitor | 750 | 2.55 | 1.39 | 0 | 8 |
| Red Cards Home | 750 | 0.09 | 0.30 | 0 | 2 |
| Red Cards Visitor | 750 | 0.08 | 0.31 | 0 | 3 |
| Total Player Substitutions | 750 | 4.49 | 1.06 | 0 | 6 |
| Attendance (1000s people) | 750 | 27.84 | 17.78 | 5.17 | 98 |
| Attendance/Capacity | 750 | 0.74 | 0.17 | 0.19 | 1 |
| Distance Home-Visitor (1,000 km) | 750 | 0.73 | 0.60 | 0 | 2.70 |

the home or the visiting team that is ahead in the score. This is not the case for games where the difference is 1 goal. When the home team is ahead by 1 goal (+1 in figure 7.1), the referee allows almost 30% less additional time than the average, whereas if the home team is behind by 1 goal (–1 in the figure), the referee allows 35% more time than the average. In both cases, the difference is statistically significant.

This phenomenon constitutes prima facie evidence of favoritism on the part of the referee. Injury time appears to systematically benefit the home team, but only in the cases of games close enough for this additional time to have a chance to matter. This observation leads us to suspect that referees may consistently favor teams simply because they play at home.

But suggestive circumstantial or prima facie evidence is far from sufficient to move beyond suspicion. If the difference in injury time does not arise through bias, what else could be responsible for the observed patterns? Law 7 in the official *Laws of the Game* (FIFA 2012) states that "allowance for injury time is made in either period of play for all time lost through substitutions, assessment of injury to players, removal of injured players for treatment, wasting time, or any other cause." Thus, perhaps "true" injury time is correlated with the *identity* of the team

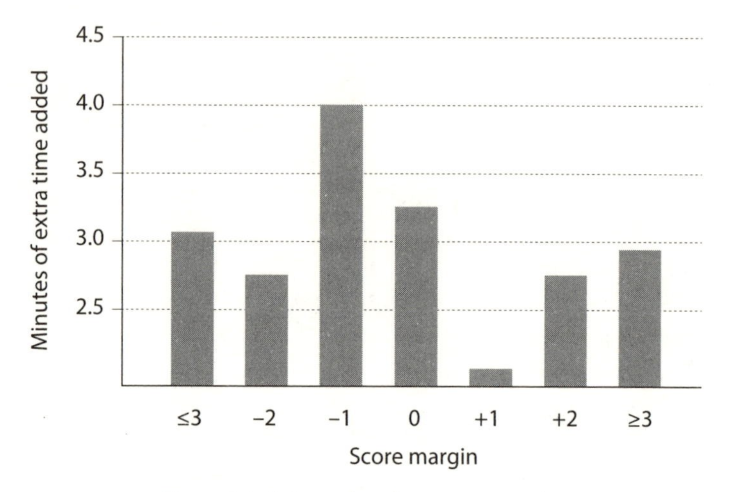


**Figure 7.1.** Injury time by score margin.

leading at the end of the game, but only in close games. To test for this, we first study whether allowing for variables correlated with the intensity of the game affects the results (see table 7.2). In particular, we estimate how our measure of bias is affected by controlling for both the numbers of yellow and red cards and the number of player substitutions.

Favoritism is captured by the coefficient on the Score Difference dummy variable, which equals 1 if the home team is ahead by 1 goal and 0 if the home team is behind by 1 goal. The univariate regression shows that on average the injury time is shorter by 1.88 minutes when the home team is ahead by 1 goal. The second specification includes controls for yellow and red cards, and the number of players replaced by a substitute. We find positive and significant effects of both yellow cards and the number of player substitutions on the amount of injury time. In other words, injury time is affected by the intensity of the game (in light of the purpose of injury time, this outcome should of course be no surprise). Interestingly, the effect of Score Difference remains stable and highly significant after including these variables. This result supports our identification strategy, in that Score Difference is not capturing the effect of game intensity on "true" injury time.

However, perhaps it is not observable differences in intensity, but rather differences correlated with the identity of the teams, that cause variation in the amount of warranted injury time. To examine this possibility, we control for the relative strengths of both teams (as measured by their ranks and operating budgets), the absolute value of the difference in the ranks, and team fixed effects. We also control for referee fixed

**Table 7.2.** Minutes of Injury Time at the End of Match in Close Games

| | | | | | | |
|---|---|---|---|---|---|---|
| Constant | 3.98** | 2.94** | 3.23** | 3.28** | 3.01** | 3.05** |
| | (0.09) | (0.17) | (0.33) | (0.60) | (0.44) | (0.70) |
| Score Difference | -1.88** | -1.86** | -1.78** | -1.77** | -1.76** | -1.80** |
| | (0.12) | (0.11) | (0.11) | (0.12) | (0.12) | (0.13) |
| Yellow Cards | | 0.08** | 0.06** | 0.05** | 0.06 | 0.06* |
| | | (0.02) | (0.02) | (0.03) | (0.03) | (0.03) |
| Red Cards | | -0.20 | -0.19 | -0.17 | -0.16 | -0.22 |
| | | (0.13) | (0.12) | (0.13) | (0.13) | (0.15) |
| Player Substitutions | | 0.14** | 0.04 | 0.04 | 0.02 | 0.08 |
| | | (0.05) | (0.07) | (0.07) | (0.07) | (0.08) |
| Year Effect | | | 0.11 | -0.09 | 0.52 | -0.10 |
| | | | (0.19) | (0.37) | (0.37) | (0.43) |
| Budget Home Team | | | 0.00 | 0.06 | -0.01 | 0.04 |
| | | | (0.02) | (0.10) | (0.02) | (0.11) |
| Budget Visiting Team | | | 0.05** | 0.05** | -0.02 | 0.06** |
| | | | (0.02) | (0.02) | (0.08) | (0.02) |
| Rank Home | | | 0.02 | 0.01 | 0.02 | 0.02 |
| | | | (0.01) | (0.03) | (0.01) | (0.04) |
| Difference in Ranks | | | -0.03* | -0.03** | -0.02* | -0.03** |
| | | | (0.01) | (0.01) | (0.01) | (0.01) |
| Team Fixed Effects? | No | No | No | Home | Visitor | Home |
| Referee Fixed Effects? | No | No | No | No | No | Yes |
| $R^2$ | 0.48 | 0.52 | 0.56 | 0.60 | 0.61 | 0.64 |
| $N$ | 268 | 268 | 268 | 268 | 268 | 268 |

*Notes*: The dependent variable is injury time at the end of the second half in games with a 1-goal difference. Standard errors in parenthesis. * and ** indicate significant at the 1% and 5% levels, respectively.

effects. The results show that when the *intensity* of the match *increases*, *more* injury time is added. In particular, when the visiting team is stronger (as indicated by a greater budget) and when the difference in rank between the visiting and the home team is smaller, the amount of injury time added is greater. Interestingly, though the percentage of the variation of injury time explained in the regressions increases substantially (from 48% in the simplest model to 64% in the most complete specification), the regression coefficient does not significantly change, neither in size nor in significance. This result corroborates our initial hypothesis.

Thus far, we have focused on injury time at the end of the second half of close matches. We consider two falsification tests by studying situations in which we would expect to find *no* evidence of the score on actual injury time. First note that referees also add injury time at the end of the first half of play. But because there are another 45 minutes to play in the second half, the marginal effect of adding one extra minute or two in the first half on the ultimate score is likely to be extremely low. Accordingly, we would expect to see little or no evidence of favoritism at halftime. Consistent with this premise, the sign of the first-half Score Difference variable in panel A of table 7.3, though positive, is of small magnitude and statistically insignificant. There is no favoritism in the first half.

A second falsification test in panel B studies another situation in which the marginal effect of adding extra time on the ultimate score is

**Table 7.3.** Falsification Tests

| | Panel A: Halftime Effects | | | Panel B: 2-Goal Difference | | |
|---|---|---|---|---|---|---|
| Constant | 0.70** | 0.78** | 1.42** | 2.76** | 2.36** | 1.42** |
| | (0.06) | (0.13) | (0.45) | (0.13) | (0.43) | (0.45) |
| Score Difference | 0.13 | 0.11 | 0.08 | -0.21 | -0.15 | -0.03 |
| | (0.08) | (0.09) | (0.09) | (0.16) | (0.16) | (0.17) |
| Yellow Cards | | -0.06 | -0.05 | | 0.06 | 0.03 |
| | | (0.03) | (0.03) | | (0.03) | (0.04) |
| Red Cards | | -0.24 | -0.18 | | 0.12 | 0.15 |
| | | (0.26) | (0.27) | | (0.15) | (0.15) |
| Player Substitutions | | 0.12 | 0.13 | | -0.04 | -0.02 |
| | | (0.08) | (0.08) | | (0.09) | (0.09) |
| Year Effect | | -0.02 | 0.17 | | 0.42 | 0.70 |
| | | (0.12) | (0.25) | | (0.24) | (0.44) |
| Budget Home Team | | 0.01 | -0.07 | | -0.03 | -0.04 |
| | | (0.02) | (0.07) | | (0.03) | (0.03) |
| Budget Visiting Team | | 0.01 | 0.01 | | 0.03 | -0.05 |
| | | (0.01) | (0.01) | | (0.03) | (0.11) |
| Rank Home | | 0.00 | -0.03 | | 0.02 | 0.01 |
| | | (0.01) | (0.03) | | (0.02) | (0.02) |
| Difference in Ranks | | 0.00 | 0.00 | | -0.01 | 0.00 |
| | | (0.01) | (0.01) | | (0.02) | 0.02 |
| Team Fixed Effects? | No | No | Home | No | No | Home |
| $R^2$ | 0.01 | 0.03 | 0.10 | 0.01 | 0.14 | 0.31 |
| $N$ | 332 | 290 | 290 | 161 | 161 | 161 |

*Notes*: The dependent variable in panel A is the time granted in the first half by the referee in games that at halftime had a 1-goal difference. In panel B, the dependent variable is injury time at the end of the second half in games with a 2-goal difference. Standard errors in parentheses. * and ** indicate significant at the 1% and 5% levels, respectively.

likely to be low: when there is a 2-goal difference in the score at the end of the second period of play. Again, the coefficient of Score Difference is of small magnitude and statistically insignificant.

These two tests provide further evidence that referees conform to the pressure from the crowd.[3] However, the premise is not simply that referees favor home teams, but instead that they are more likely to do so when the returns to satisfying the crowd are greater. Therefore, it is important to study situations where the preferences of the crowd may induce changes in a referee's behavior in predictable ways.

First, let us examine what happens to referee bias when the rewards of winning for the crowd are changed exogenously. As mentioned already, rewards from winning changed from 2 to 3 points after the 1994–95 season. Consider then the case where the home team is behind by 1 goal. If the home team scores, it gains 1 point under both regimes. But if the home team is ahead by 1 goal, the marginal return to finishing the game for the home team increases from 1 point to 2 points (if they concede a goal, they previously went from 2 points to 1; now they go from 3 points to 1). This exogenous variation in rewards allows us to test whether referees respond to the desires of the home crowd. This test can be done by including interaction terms between the year of observation and the Score Difference dummy, and the hypothesis is that the size of the coefficient on Score Difference increases after the points change. Consistent with this prediction, the interaction is negative and significant, which implies that the bias is stronger after the points change (see table 7.4). In numerical terms, the 1994–95 season saw a difference of 1 minute and 30 seconds, which increased to almost 2 minutes by the 1998–99 season.

Second, it is also possible to exploit the exogenous variation in *perceived* importance of the matches caused by their relative closeness to the end of the season. Teams care about their final position in the league table. Therefore, games at the end of the season may be deemed more important than those earlier in the season, both because the end of the season is closer and because teams have a better idea of their likely finishing position. Whether or not later games are actually more important does not matter, as long as stadium crowds perceive it that way and thus are more vocal in their support the nearer the prize of winning trophies (or avoiding relegation) draws. To test for this, we study how the amount

3 Garicano et al. (2005) consider an additional test: how referees respond to goals *in extra time* as a function of who scores. Consider a game that is a draw. If the home team scores, a referee who is biased in favor of the home team has an incentive to quickly signal the end to the game, whereas if the away team scores, the referee is more likely to extend the game in the hope that the home team can respond. Consistent with this intuition, when the visiting team scores, the amount of injury time is significantly greater, by roughly 20% of the average injury time.

**Table 7.4.** Marginal Effect of Incentives on Injury Time

| | | | | |
|---|---|---|---|---|
| Constant | 3.50** | 3.11** | 2.93** | 2.42** |
| | (0.14) | (0.32) | (0.34) | (0.39) |
| Score Difference | -1.53** | -1.56** | -1.47** | -0.64* |
| | (0.18) | (0.18) | (0.17) | (0.28) |
| Year Effect | 0.81** | 0.70** | 0.49 | 0.55 |
| | (0.18) | (0.21) | (0.25) | (0.27) |
| Year × Score Difference | -0.58* | -0.52* | -0.51* | -0.55* |
| | (0.23) | (0.23) | (0.23) | (0.23) |
| Yellow Cards | | 0.07** | 0.06* | 0.06* |
| | | (0.02) | (0.02) | (0.02) |
| Red Cards | | -0.20 | -0.19 | -0.09 |
| | | (0.13) | (0.12) | (0.12) |
| Player Substitutions | | 0.03 | 0.05 | 0.04 |
| | | (0.07) | (0.07) | (0.07) |
| Budget Home Team | | | -0.01 | -0.02 |
| | | | (0.02) | (0.02) |
| Budget Visiting Team | | | 0.05** | 0.04* |
| | | | (0.02) | (0.02) |
| Rank Home | | | 0.01 | 0.01 |
| | | | (0.01) | (0.01) |
| Difference in Ranks | | | -0.03** | -0.04** |
| | | | (0.01) | (0.01) |
| Game Number | | | 0.01 | 0.02** |
| | | | (0.01) | (0.01) |
| Game Number × Score Difference | | | | -0.02** |
| | | | | (0.01) |
| $R^2$ | 0.5678 | 0.5802 | 0.6107 | 0.6438 |
| $N$ | 268 | 268 | 268 | 268 |

*Notes*: The dependent variable is the time granted in the second half by the referee in games with a 1-goal difference. Standard errors in parentheses. * and ** indicate significant at the 1% and 5% levels, respectively.

of bias depends on Game Number. This variable runs from 1 (the first game of the season) to 38 (the final game of the season). We find, first, that the coefficient on Score Difference remains unchanged when we control for the stage of the season (column 3). Second, when we also interact Game Number with Score Difference (column 4), we find that the referee bias does indeed *increase* as the season advances. From the beginning to the end of the season, the referee bias increases by about 40 seconds for the −1 relative to the +1 matches.

**Table 7.5.** Effect of Size and Composition of the Crowd on Referee Bias

| | | | | |
|---|---|---|---|---|
| Constant | 3.23** | 2.94** | 2.65** | 4.09** |
| | (0.18) | (0.20) | (0.26) | (0.44) |
| Score Difference | -0.93** | -0.96** | -0.88** | -2.92** |
| | (0.20) | (0.21) | (0.20) | (0.47) |
| Year Effect | 0.36** | 0.33** | 0.12 | 0.12 |
| | (0.11) | (0.11) | (0.18) | (0.18) |
| Attendance | 0.00 | 0.00 | 0.01 | 0.01 |
| | (0.00) | (0.00) | (0.01) | (0.01) |
| Attendance x Score Difference | -0.02** | -0.02** | -0.02** | -0.02** |
| | (0.00) | (0.00) | (0.00) | (0.00) |
| Yellow Cards | | 0.07** | 0.05* | 0.05* |
| | | (0.02) | (0.02) | (0.02) |
| Budget Home Team | | | 0.00 | 0.00 |
| | | | (0.04) | (0.04) |
| Budget Visiting Team | | | 0.05* | 0.05** |
| | | | (0.02) | (0.02) |
| Rank Home Team | | | 0.02* | 0.02 |
| | | | (0.01) | (0.01) |
| Difference in Ranks | | | -0.03* | -0.02 |
| | | | (0.01) | (0.01) |
| Game Number | | | 0.01 | 0.01 |
| | | | (0.00) | (0.00) |
| Ratio of Attendance to Capacity | | | | -0.51 |
| | | | | (0.37) |
| Ratio of Attendance to Capacity × Score Difference | | | | 1.51** |
| | | | | (0.32) |
| $R^2$ | 0.5678 | 0.5802 | 0.6107 | 0.6438 |
| $N$ | 255 | 255 | 255 | 255 |

*Notes*: The dependent variable is the time granted in the second half by the referee. The effect of the crows is given by the interactions involving Attendance. Standard errors in parentheses. * and ** indicate significant at the 1% and 5% levels, respectively.

A third and final test notes that if in fact stadium audiences pressure referees into favoring their preferred team, then the size of the crowd and its composition should matter for the amount of referee bias (see table 7.5).

First, we can look at the relationship between crowd size and injury time. On average, attendance does not seem to significantly affect the amount of injury time added. However, on the margin (comparing the +1 and −1 situations that are interacted with Score Difference), it does:

A one-standard deviation increase in crowd size increases the bias by approximately 20%. This effect, however, arises predominantly from the larger stadiums of more popular teams. This result means that econometrically it is not possible to distinguish between attendance and home team effects.

However, we can examine *unusually* large audiences on the bias shown by the referee by looking at the ratio of crowd size to stadium capacity. Therefore, we would like to test whether referees are likely to be less biased in favor of the home team when attendance is unusually high (and, therefore, typically a higher proportion of fans supports the away team).[4] Consistent with the hypothesis, the effect of unusually high attendance interacted with the Score Difference is highly significant and results, as predicted, in less bias.

Two final small results, perhaps more for soccer aficionados. The referees' susceptibility to social pressure is quite homogeneous in that most referees appear to be equally biased. Only 3 of the 35 referees in the sample show statistically significant individual effects (at the 10% level). The referees' favoritism is also quite homogeneous across teams. Two teams, however, particularly benefit from the referees' bias: Barcelona and Real Madrid. They have statistically significant individual effects, which means that they significantly receive more bias in favor and less bias against than any other teams. On this second result, I suspect that most soccer fans who follow La Liga would claim that one does not need an econometric regression to know this.

To conclude, let us address some alternative hypotheses and briefly touch upon some newer contributions in this field.

First, we can be reasonably sure that the differences in the extra time that is added cannot be explained by the intensity of the game. Including measures of intensity leaves the size and significance of the effect intact. But how about differences in the strategies that the teams follow? The strategies may potentially differ widely in close matches. To further study this point, in addition to the available controls studied earlier, we have also looked at the time the leading goal is scored. For the favoritism hypothesis it should make no difference, but if the amount of extra time reflects strategic behavior by the teams (e.g., time wasting), then it

4 Garicano et al. (2001) find that large crowd size relative to capacity can be explained by popular teams visiting (Barcelona or Real Madrid, the only ones with official supporters clubs in every province in Spain) and by geographically close teams playing each other. This phenomenon means that unusually large crowds relative to average are indicative of a higher proportion of the crowd supporting the away team.

should make a difference. Consistent with the favoritism hypothesis, the size and significance of the effect remain basically intact.

Second, an alternative hypothesis to the idea of social pressure is that instead referees take bribes. This hypothesis, however, is unconvincing: There is no apparent reason to believe that the ability to bribe depends on whether a team is playing at home or away. One hypothesis that we cannot rule out is the possibility that the governing body, the Real Federación Española de Fútbol (RFEF), condones this form of favoritism. The evidence shows that crowd preferences affect referee behavior, but it might be the case that RFEF (the principal) instructs the referee (the agent) to take crowd preferences into account. Although this idea is perhaps intuitively appealing, it is not likely that authorities systematically favor home versus away teams, for a number of reasons. One is that the international governing body FIFA has shown its disapproval of such referee behavior by changing the rules in 1998 to force them to commit ex ante to the amount of injury time. Another reason is that though the RFEF may favor some teams over others, it seems implausible that it would systematically favor home over away teams. And, finally, perhaps the RFEF would like close games to continue longer (as these are most exciting), but why then are these games shorter when the home team is ahead?

Although the shouting, whistling, and singing of the crowd may often give a different impression, it turns out that the actual number of games affected by this type of referee bias is small. The estimates suggest the result of approximately 2.5% of all the games (about seven games) in the sample changed because of it. However, though this is the one form of referee bias we can empirically verify, it is unlikely the only one. Other forms may include the subjective interpretation in favor of the home team of fouls, offside decisions, penalties, and other rules. Therefore, the estimates obtained here could be seen as a lower bound on the favoritism shown by referees.

*

Sir Stanley Ford Rous (April 25, 1895–July 18, 1986), Order of the British Empire, and 6th President of FIFA from 1961 to 1974, served as secretary of the Football Association in England from 1934 to 1962. He also was an international referee, and in a lecture he gave in 1969 he noted the following:

> Referees are basically honest and impartial, but they do react differently to situations. How many referees will give a penalty against the

home team early in the match, when play is often most fierce? We have all seen indirect free-kicks given in the penalty area instead of a penalty kick for one of the nine penalty offences. We have all seen referees whistle for penalty offences inside the area, then place the ball a foot or so outside the area. Thus degrees of punishment, instead of correct disciplinary action, are being applied. . . . In an international tournament recently, I saw a referee give a penalty—in my view a harsh decision. The players, including the goalkeeper, took up proper positions without appeal, and the player taking the kick shot the ball straight into the goalkeeper's hands. He was ordered to retake the kick and scored—an absolute gift from the referee! At the "inquest," the referee said that he was not ready for the kick to be taken and that he had not blown his whistle. At a European referees' conference two years back, when I was making the point that an offence must be punished regardless of the score and at any time in the match, I was shocked when one of the most famous and experience referees of the time said, "That's all very well, but I would never give a penalty against Austria in Vienna during the last few minutes of a match—and hope to get away safely!" The younger referees present were astonished at this confession.

Physical violence was an important aspect of refereeing in the 1960s and 1970s. It is highly unlikely, though, that the empirical results are the outcome of referees being afraid of physical violence from the crowd. Physical violence has become exceedingly rare, to the point that the fences that were erected in Spanish stadiums in the 1970s as a precaution against violence were taken away in the early 1990s. The same happened across European leagues.

After the pioneering work on the effects of nonmonetary forces on behavior discussed in this chapter, a number of authors have followed similar paths. In soccer, the study by Petterson-Lidbom and Priks (2010) is particularly interesting. When the Italian government forced teams with deficient security standards to play their home games without *any* spectactors, the advantage in terms of the normal foul rate, yellow cards, and red cards typically afforded to home teams disappeared entirely.[5] At the other extreme, after the deaths and arrests that followed the sheer terror that hooligan violence (see chapter 9) generated in Argentina in the 2012–13 season, as a matter of national security the Argentinian government is forcing teams to play all their games in the 2013–14 season with *only* home team supporters. Supporters of the visiting team are not allowed in stadiums in league games. It would be interesting to study

5 See also Rickman and Witt (2008) and Dawson and Dobson (2010).

the effects on the form of favoritism studied in this chapter. For other sports, Moskowitz and Wertheim (2011) discuss a number of academic studies showing not only that the referee bias from social influence is also present in professional US leagues, such as the NBA (basketball), MLB (baseball), and NHL (hockey), but that it is the *leading cause of home field advantage* in these leagues.

# 8

# MAKING THE BEAUTIFUL GAME A BIT LESS BEAUTIFUL

(with Luis Garicano)

I do not, however, deny that I planned sabotage. I did not plan it in a spirit of recklessness, nor because I have any love of violence. I planned it as a result of calm and sober assessment.

—NELSON MANDELA'S STATEMENT AT THE OPENING OF THE DEFENSE CASE IN THE RIVONIA TRIAL (PRETORIA SUPREME COURT, APRIL 20, 1964)

STRONG INCENTIVES OFTEN HAVE DYSFUNCTIONAL CONSEQUENCES. CIA FIELD agents rewarded on the number of spies recruited fail to invest in developing high-quality spies (WMD Commission Report 2005, p. 159). Civil servants rewarded on outcomes in training programs screen out those who may most need the program (Anderson et al. 1993; and Cragg 1997). Training agencies manipulate the timing of their trainees' performance outcomes to maximize their incentive awards (Courty and Marschke 2004). Teachers cheat when schools are rewarded on student test scores (Jacob and Levitt 2003). A theoretical literature going back at least 30 years (for instance, Kerr 1975; Holmstrom and

Milgrom 1991; and Baker 1992) has studied the possibility of dysfunctional responses to incentives in different settings. Essentially, as Baker (1992) carefully argues, when output is not clearly observed, what matters is the correlation, on the margin, between what is rewarded and the desired action.

Dysfunctional responses may occur not only in cases of individual incentive contracts but also in settings where individuals compete with each other and are rewarded on the basis of relative performance. In these settings, strong incentives may be particularly damaging if agents can devote resources not only to productive activities but also to depressing each other's output.

However, whereas anecdotal accounts of "back-stabbing," badmouthing, and other sabotage activities are easy to find, there does not exist any systematic work documenting such responses. An obvious reason why such actions are usually *impossible* to document is that workers who sabotage their fellow workers' performance typically go to great lengths to conceal their actions.

Viewed from this perspective, this chapter studies an incentive change in a natural setting where both productive and sabotage activities can be directly observed. The setting is, not surprisingly, the most popular sport in the world: soccer.

As we know from the previous chapter, soccer teams that engage in league competition (round-robin tournaments) were historically awarded 2 points for winning a match, 1 point for tying, and no points for losing. In the run-up to the football World Cup that was to take place in the United States in 1994, however, FIFA decided to change the reward for the winning team from 2 points to 3 points while leaving the reward for ties and losses unchanged.

The objective of FIFA, worried about the possibility of empty stadiums in the United States, was to raise the incentive to attack in games, with a view to driving up the number of goals and overall excitement levels (for example, *USA Today* 1994). The *Los Angeles Times* (Dwyre 1993) reports: "An underlying reason for FIFA's action, and for World Cup Chairman Alan Rothenberg of the United States pushing hard for it, was the feeling that American fans, used to higher-scoring American games, would be much less tolerant and much more quickly turned off than a more traditional soccer audience by an early parade of 0–0 and 1–1 results."

Citing experts of the game, *The New York Times* (Yannis 1994) commented on the decision: "A decision by FIFA last June to reward teams three points for a first-round victory instead of two has increased optimism that teams will emphasize offense and produce a scoring spectacle in the World Cup."

This change subsequently became part of the *Laws of the Game* (FIFA 2012) and was applied after 1995 to all league competitions worldwide.[1] Interestingly, little if any intellectual analysis about the potential effects of the rule change was done along the way.

We use a detailed data set on football matches in Spain before and after the change to study the effect of this change in rewards along a number of dimensions. In this context, we call "sabotage" any effort that is intended to reduce the performance of the rival in the match. In particular, we focus our analysis on all such destructive actions that are perceived as "dirty play" or "negative play" and penalized in different ways in the *Laws of the Game*.

Our setting has two key advantages. First, negative activities are *observable*. We have information on the type of specialists in different actions (productive and destructive) that teams choose to field. More importantly, both productive actions aimed at increasing one's own output and destructive actions aimed at decreasing the opponents' output are observed and routinely recorded in newspapers and box scores. Second, we can take advantage of an unusual control group: The same teams that engage in league play were playing at the same time in a cup tournament that experienced no changes in incentives. Using their behavior in this tournament, we can eliminate the effect of any changes in styles of play or other time trends unrelated to the incentive change.

The change to the three-point rule that we study should lead teams to try harder to win. This attempt to win may result in two types of actions: Teams may undertake more offensive actions, but they may also play "dirtier" (unsporting behavior punished in different ways) because it now becomes more important to prevent the opposing team from scoring a goal. Stronger incentives may then lead to more negative play. For example, tackling an opponent may reduce his or her likelihood of scoring but also poses an important physical risk to both players. An increase in the value of winning may thus lead to an increase in this type of effort. Does then the amount of dirty play increase? And if so, is it possible to say that this is "bad," and therefore unintended, as opposed to providing simply a more intense, and perhaps even more fun, game? Put differently, are stronger incentives detrimental to the objective of FIFA?

Our analysis proceeds in four steps, as follows. First, we start by describing the basic behavioral changes that took place after the rule change. We find that, consistent with what we might expect, the introduction of the new incentives was followed by a decrease in the number of ties. However, the number of matches decided by a large number of

1 Professional soccer leagues in England had already introduced this change in the reward schedule in 1981, that is, beginning in the 1981–82 season.

goals declined. Measures of offensive effort, such as shot attempts on goal and corner kicks, increased, while indicators of sabotage activity, such as fouls and unsporting behavior punished with yellow cards, also increased after the change. Of course, all of these results could follow simply from time trends and, hence, they are merely suggestive at this point.

Second, we proceed to use the control matches in the cup tournament to estimate the effects caused by the change in rewards. Most, but not all, of the changes we observe in the previous before–after analysis are still present in the differences-in-differences (DID) analysis we implement. We observe an increase on the order of 10% in the measures of attacking effort desired by FIFA. We find, however, that the number of fouls increased significantly, by around 12.5%, as a result of the incentive change. The net result of these opposing forces is that the number of goals scored did not change.

We then try to understand the underlying mechanisms through which these changes took place and the reason they neutralized each other in terms of goal scoring by examining the way the behavior of teams changed *during* the match. We expect teams that get ahead by one goal to become more conservative, since conceding one goal from this position would cause them now to drop two points rather than one point. On the other hand, the behavior of teams that get behind should not change a great deal because the marginal value of one goal (tying) remains basically unchanged.[2]

The evidence we find is consistent with this hypothesis: Teams that get ahead become more conservative by increasing significantly the number of defenders they use. This change in the defensive stance has two consequences: The probability of scoring an additional goal by a team that is ahead drops significantly; moreover, by the end of the match, the losing team ends up making significantly fewer *attempts* on goal than before the incentive change. Hence, the winning team successfully manages to "freeze the score."

The fourth and final step is actually to show that this change represented *undesirable* sabotage rather than, say, desirable greater intensity in the games. That is, we try to understand the welfare consequences of the stronger incentives that are implemented. Public statements by FIFA officials indicated that, in the spirit of Kerr (1975), they were increasing the rewards for wins while hoping for more scoring; this result, we know, did not happen. Still, a more intense match could be more fun

2 Under the new incentive scheme, the reward for a tie (one point) is a lower proportion of points per win. On the other hand, there is an increase in the value of scoring one goal on the way to scoring two, in terms of the option it gives on winning the match.

even without more goals, if the public likes the greater emphasis on defense. We find that this was not the case either. We exploit the lack of selection in the assignment of teams to stadiums given that all teams play in all stadiums and calculate the effect of playing at one's home stadium against a "dirtier" team, measured in several different ways. Controlling for team fixed effects, we find that attendance at any given stadium *decreases* significantly when the stadium is visited by teams that play dirtier. This result is important in that it confirms the idea that the significant increase in sabotage actions we find is, on the margin, undesired by the public. We finally show that, indeed, attendance at stadiums decreased as a result of the sabotage.

We conclude this chapter with a brief discussion of the potential relevance our findings have for agency problems and the tournaments literature. Based on our findings, we also discuss how teams might respond to recent proposals to change other rules. Thus far, rule changes have been discussed and decided on with little data and even less data analysis. From this perspective, our relatively speculative analysis may represent a contribution to this discussion. Overall, the evidence suggests, consistently with the broad empirical agency theory literature (see Gibbons 1998 and Prendergast 1999, 2002 for reviews) that soccer clubs reoptimized and changed their behavior in response to stronger incentives but that they did this largely in a manner undesired by the principal: They engaged in more sabotage activities and managed to decrease the output desired by the principal. The beautiful game became a bit less beautiful. Thus, we see our evidence as supporting incentive models with multiple tasks, where the cost of increasing incentives is encouraging more effort of the "wrong" kind.

*

The data were obtained from *Marca*, which is the best-selling newspaper in Spain, and from www.sportec.es. The setting concerns the Spanish League competition La Liga, and we use data from the 1994–95 full season (370 games), the last one with the 2–1–0 scheme, and from the 1998–99 full season (380 games) with the new 3–1–0 scheme. Using data that are four seasons apart is convenient because, as in the previous chapter, it does not require us to assume that teams immediately adjust their behavior to the new situation. It also means that we will have to account for any possible year effects in the data. To do this, we use data from the Spanish Cup competition *Copa del Rey* as controls in our analysis. This competition is an elimination tournament in which teams are randomly paired together, no points are awarded, and the winner survives to the next round. All changes in rules and regulations that took place during the period of analysis apply equally to league and cup games *except*, of course, the change in rewards in league games. As a result, the behavior of the teams in the cup tournament should be largely unaffected by the change in the reward scheme in the league tournament.[3]

We have obtained detailed observations of multiple measures of actions, both sabotage and the desired attacking or offensive effort, along with the teams' choices of specialists. They are described as follows.

## PLAYER TYPES

In a soccer game, each team lines up one goalkeeper and ten field players. Field players can be of three possible types: defenders, midfielders, or attackers. Defenders, who play closest to their own goal, defend it when it is under attack. This play often requires stopping rival players through hard tackles or other types of dirty play. Thus, they are most likely to be involved in sabotage activities. Attackers, or forwards, are the primary scorers who play closest to the other team's goal. They are players specialized in the type of effort (attacking actions) that FIFA wants to increase.[4] Lastly, midfielders play between defenders and attackers, and their role is to support both of these types of players.

We classify each of the players in every team that played in every match in the sample using the official classification of players' types published by *Marca* and www.sportec.es. The data include information on the number of the different types of players at the beginning of each match and *during* each match. Although our main direct evidence comes from changes in observed actions, the information on player types is useful to study teams' defensive and attacking stances.

## ACTIONS

For every match and for every team in the sample, the data set includes information on the number of destructive and productive actions.

3 If anything, this control group of games provides us with a lower bound on the effects of the change in incentives. The reason is that players may adapt their style of play to the new reward scheme in the league and, as a result, change how they play in *both* league and cup games. We use two years of cup data before and two years after the change (1993–94 and 1994–95 before and 1997–98 and 1998–99 after) to have a greater number of matches in our sample since in an elimination tournaments the number of total matches is smaller. We have also checked that the chosen years are not outliers in terms of average goals scored, fouls, and other variables in league matches relative to cup matches.

4 Data from *Marca* (2012) show that indeed sabotage actions are committed mainly by defenders and attacking actions mainly by attackers (e.g., more than two-thirds of all fouls and yellow cards are given to defenders, and attackers represent more than 70% of the players who score at least one goal).

## Destructive Actions

### Fouls

In the *Laws of the Game* (FIFA 2012), the following actions are sanctioned as fouls: "Tripping or attempting to trip an opponent, charging into an opponent, striking or attempting to strike an opponent, pushing an opponent, jumping at an opponent in a careless or reckless manner or using excessive force, blatant holding or pulling an opponent, and impeding the progress of an opponent." These actions are penalized in different ways.[5]

In addition to fouls, there are two color "cards" that the referee holds up to indicate hard fouls and behavior that will not be tolerated: yellow cards and red cards.

### Yellow Cards

Yellow cards indicate a formal "caution" for any form of "unsporting behavior," which includes especially "hard fouls, harassment, blatant cases of holding and pulling an opposing player, persistently breaking the rules," and other similar acts (FIFA 2012). In addition to being punished as a foul, a player who receives two yellow cards is given a red card and ejected from the game without being replaced by a teammate.

### Red Cards

Red cards are given after a second yellow card is given in the same match, as well as for behavior that is clearly beyond the bounds of the game such as "violent conduct, spitting at an opponent, using offensive or threatening language, and use of excessive force or brutality against an opponent."

It seems apparent that these three types of destructive actions (fouls, yellow cards, and red cards) are aimed at reducing the rivals' output. Empirically, around 85–90% of all such sabotage activities are fouls where players are not booked with a card, 10–15% are fouls where a yellow card is given, and typically less than 1% are actions punished with a red card. For the most part, we focus our attention on fouls and yellow cards.

## Productive Actions

With regard to actions aimed at scoring, we have data on shots, which are attempts on the opposition team's goal that missed the target, and shots on goal, which are those that did not miss the target. The data also include corner kicks, an action that is a consequence of attacking

5 Depending on the action and its severity, they are punished with either a direct free kick or an indirect free kick. If they take place inside the penalty box, they are punished with a penalty kick. See Law 12 on fouls and misconduct in FIFA (2012).

behavior: If during an attack the ball goes out of bounds over the end line and was last touched by the defending team (e.g., a shot that was deflected by a defender), the attacking team inbounds it from the nearest corner by kicking it in from the corner arc.

## OTHER VARIABLES

We also have data on the date of the game, the stage of the season (game number), the winning record of each team at the time of the match, stadiums' capacities, attendance at each match, and the operating budgets of each team, a proxy for the strength of a team. Lastly, our data set includes the number of goals by each team and their timing, as well as information on extra time or injury time and player substitutions:

### Extra Time or Injury Time

As indicated in the previous chapter, soccer games have two 45-minute halves, at the end of which the referee may, at his or her discretion, award what is often referred to as "extra time" or "injury time." Law 7 in the official *Laws of the Game* states that "allowance for injury time is made in either period of play for all time lost through substitutions, assessment of injury to players, removal of injured players for treatment, wasting time, or any other cause. Allowance for time lost is at the discretion of the referee" (FIFA 2012). Information on the amount of extra time that referees add on may thus be valuable as indirect, additional evidence on the amount of destructive actions that took place.

### Player Substitutions

Players may be replaced by a substitute at any time during the match. Teams may use up to a maximum of three substitutes. We have information on the timing at which substitutions take place.

We begin in figure 8.1 by presenting the probability distribution of score margins before and after the change. The percentage of all matches that ended in a tie decreases from 29.7% to 25.5%, and the number of matches decided by a single goal (whether in favor of the home or visiting team) experiences a large increase, from 31% to 40%. In absolute terms, the number of tied games decreased from 110 to 97, the number of matches that finished with a 1-goal difference increased from 115 to 153, and those that finished with a difference of two goals or more decreased from 145 to 130. Statistically, the before and after distributions are significantly different (Pearson $\chi^2(6) = 17.28$; $p$-value: 0.008).[6]

6 We omitted margins above three games to conform to the practice of limiting the Pearson analysis to bins for which the expected number of observations is at least five.

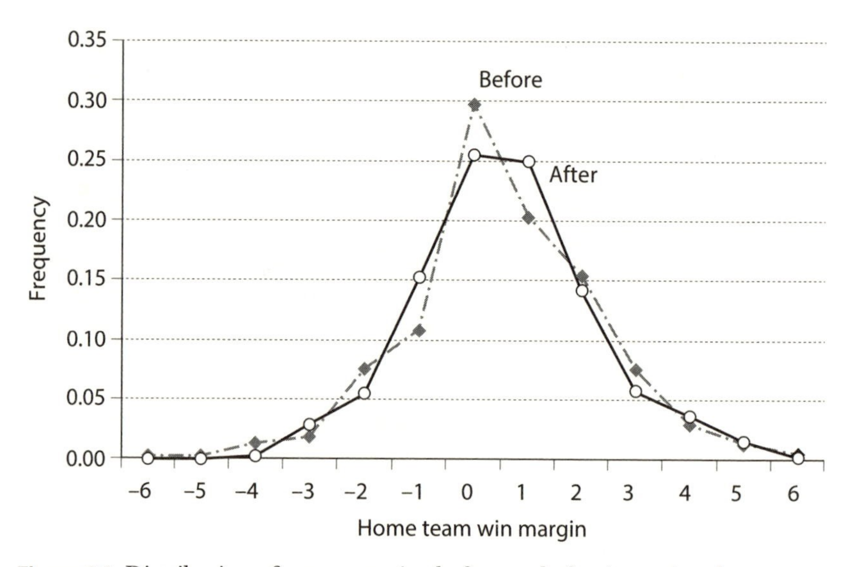


**Figure 8.1.** Distribution of score margins before and after incentive change.

This first look at the data, therefore, suggests a clear, nonmonotonic pattern in the outcomes: Teams are less likely to tie, but they are also less likely to win by a typically "useless," but possibly quite entertaining, large number of goals.

Table 8.1 presents some descriptive statistics before and after the change. This table does not account for possible year effects, as it only reports changes in means, but it gives an idea of the main patterns observable in the data. It also shows that the effects that we find in the next section result, as we might expect, from changes in the "treatment group." We find, for instance, that there were statistically significant and large increases in regular fouls, yellow card fouls, shots on goal, and corner kicks. With respect to match outcomes, we see the drop in the proportion of ties referred to before, as well as an increase in extra time and a decrease in attendance.[7]

As indicated earlier, these results, though suggestive, could simply reflect other trends in the way soccer is being played. We proceed in the next section to study the relations of these changes to the changes in

7 Consistent with conventional wisdom, clubs play more defensively in away games. The squad composition measured by, say, Number of defenders – Number of forwards, is +1.05 (away games minus home games). This home–away difference is also affected by the three-point rule: This difference becomes +1.17, more than a 10% increase. Furthermore, there is also an induced increase in defensiveness in protection of a lead that is more pronounced at away games after the three-point rule.

**Table 8.1.** Before-After Estimates

| | Before | After | Difference |
|---|---|---|---|
| Offensive Play | | | |
| Attackers | 2.08 | 2.35 | 0.274*** |
| | (0.0244) | (0.0256) | (0.0353) |
| | $N = 740$ | $N = 760$ | |
| Shots | 6.19 | 6.80 | 0.619*** |
| | (0.124) | (0.101) | (0.16) |
| | $N = 734$ | $N = 760$ | |
| Shots on Goal | 4.12 | 4.75 | 0.626*** |
| | (0.0882) | (0.0775) | (0.117) |
| | $N = 760$ | $N = 760$ | |
| Corner Kicks | 5.29 | 5.94 | 0.649*** |
| | (0.101) | (0.0885) | (0.134) |
| | $N = 734$ | $N = 760$ | |
| Sabotage Play | | | |
| Defenders | 4.05 | 3.93 | -0.122*** |
| | (0.0286) | (0.03) | (0.0415) |
| | $N = 740$ | $N = 760$ | |
| Fouls | 16.20 | 17.49 | 1.290*** |
| | (0.191) | (0.151) | (0.243) |
| | $N = 734$ | $N = 760$ | |
| Yellow Cards | 2.33 | 2.67 | 0.338*** |
| | (0.0549) | (0.0614) | (0.0823) |
| | $N = 734$ | $N = 760$ | |
| Match Outcomes | | | |
| Goals Scored | 1.25 | 1.32 | 0.064 |
| | (0.0443) | (0.0432) | (0.0618) |
| | $N = 740$ | $N = 760$ | |
| Tied Matches | 0.297 | 0.255 | -0.042 |
| | (0.0238) | (0.0224) | (0.0327) |
| | $N = 370$ | $N = 380$ | |
| Extra Time | 3.46 | 3.97 | 0.506*** |
| | (0.0647) | (0.0593) | (0.0878) |
| | $N = 370$ | $N = 380$ | |
| Attendance | 0.755 | 0.719 | -0.035*** |
| | (0.00845) | (0.00949) | (0.0127) |
| | $N = 370$ | $N = 380$ | |

Notes: This table reports differences in offensive and defensive effort and selected match-level statistics in league soccer matches before and after the FIFA incentive change. For the offensive and defensive measures, the unit of observation is a team within a match. For the match outcomes, the unit of observation is a match except for goals; then, it is a team within a match. Attendance is measured as the fraction of available seating that was occupied. Where appropriate, standard errors, reported in parentheses, have been adjusted for clustering on match. *** denotes significant at the two-tailed 1% level.

incentives, by comparing them with the changes that took place in the cup tournament *Copa del Rey*.

## RESPONSES TO THE THREE-POINT RULE

As mentioned earlier, we consider player types as an indication of the teams' defensive and attacking stances. Changing the composition of player types, therefore, may be taken just as suggestive evidence of how teams may respond in their choices of productive and destructive effort.[8]

Direct evidence comes from changes in actions. With respect to actions, the stated purpose of the change was to encourage attacking and scoring, so attacking actions are desired by the principal per se, especially if they lead to more scoring. On the destructive side, hindering the opponent's ability to compete by injuring opposing players and other forms of dirty play punished with fouls and yellow cards seem unquestionable sabotage activities.

For each outcome variable, we first present the simple DID estimator, which is the difference of the difference in means. The effect of the incentive change is then the interaction between league (non-cup) and year. Then, we repeat the analysis controlling for the strength of the teams in the match using their operating budgets, and lastly we add team fixed effects.

### Attacking Play

Table 8.2 presents our main evidence on these types of actions. We have a number of proxies for attacking behavior:

1. Player Types. First, we find that there is a large and significant increase in the number of attackers as a result of the change, estimated at 0.41. Considering that 2.08 forwards were used on average before the change, this estimated 20% increase is in fact sizable. Controlling for the budgets of the teams (column II) or team's fixed effects (column III) reduces the coefficient estimates to about 0.28. The evidence from these three specifications is nevertheless unambiguous: Teams significantly increase, by roughly between 0.28 and 0.41 players per team, the number of attackers they use as a result of the new reward scheme.
2. Attacking Actions. We construct a proxy of offensive or "good" effort using the first principal component of three variables: corner

8 Moreover, the theoretical literature treats agents as individuals, not as teams of different types, and hence yields implications only for the *actions* that agents take as a response to incentives.

**Table 8.2.** Changes in Desired Offensive Effort

| | Attackers | | | Offensive Index | | |
|---|---|---|---|---|---|---|
| Explanatory Variable | (I) | (II) | (III) | (IV) | (V) | (VI) |
| Incentive Change | 0.413*** | 0.286* | 0.276*** | 0.287** | 0.239 | 0.256* |
| | (0.098) | (0.148) | (0.0887) | (0.133) | (0.151) | (0.142) |
| Cup Dummy | 0.12 | 0.00345 | 0.104* | −0.371*** | −0.43*** | −0.388*** |
| | (0.0759) | (0.121) | (0.0622) | (0.0937) | (0.107) | (0.104) |
| Year Effect | −0.139 | −0.0401 | −0.216*** | 0.135 | 0.0647 | 0.139 |
| | (0.0914) | (0.144) | (0.0835) | (0.12) | (0.142) | (0.132) |
| Visitor Dummy | | −0.183*** | −0.189*** | | −0.478*** | −0.473*** |
| | | (0.0347) | (0.0293) | | (0.0479) | (0.0472) |
| Own Budget | | 1.922E-5*** | | | 1.99E-5** | |
| | | (6.116E-6) | | | (8.699E-6) | |
| Opponent's Budget | | −8.74E-6 | | | 2.304E-5*** | |
| | | (5.997E-6) | | | (8.643E-6) | |
| Intercept | 2.08*** | 2.15*** | 2.28*** | −0.182*** | −0.0252 | 0.0691 |
| | (0.0244) | (0.0337) | (0.0275) | (0.042) | (0.0591) | (0.0522) |
| Include team fixed effects? | No | No | Yes | No | No | Yes |
| $N$ | 1698 | 1574 | 1698 | 1596 | 1568 | 1596 |
| $R^2$ | 0.036 | 0.062 | 0.280 | 0.050 | 0.103 | 0.134 |

*Note*: This table reports differences-in-differences estimates of the effect of a change in scoring incentives on the number of attackers initially deployed by a team and an offensive index. The offensive index is the first principal component of the number of shots, shots on goal, and corner kicks made by a team. The unit of observation is a team within a match. The first difference compares matches in seasons before and after the incentive change, and the second difference compares matches in the cup tournament to league play. Standard errors clustered on matches are reported in parentheses. * denotes significance at the 10% level, ** at the 5% level, and *** at the 1% level.

kicks, shots, and shots on goal. The results are reported in columns IV, V, and VI. We see a clear increase in offensive effort, suggesting that the incentive change resulted in an increase in the number of shots, shots on goals, and corner kicks. We also calculated the effects for the individual components of the index and, although not separately statistically significant, they all showed increases of around 10%.

## Negative Play

Table 8.3 reports the effect of the incentive changes on sabotage activities. We study three measures of sabotage:

1. Player Types. We find in columns I, II, and III that the number of specialists in defense *increases* from 0.10 to 0.25 depending on the specification. Given that the average prechange number of defenders is 4, these amounts represent an increase of about 2% to 6%. Note that this is one instance where the differences-in-differences estimates reverse the before–after findings.
2. Fouls. The second panel, columns IV, V, and VI in the table, performs the same analysis for regular fouls. Recall that this type of fouls represents the large majority of all sabotage activities. The result here is quite conclusive: The incentive change produced a precisely estimated increase in the number of fouls of about 2. Given the prechange mean of 16.2, the estimate represents approximately a 12.5% increase in the number of fouls as a consequence of the incentive change.
3. Yellow Cards. Because referees are subject to an upper limit on the number of yellow cards they can give per player (because two yellow cards to the same player in a game causes that player to be expelled), yellow cards may be less sensitive than other measures of sabotage. Consistent with this intuition, all the estimates we obtain in columns VII, VIII, and IX are positive and of comparable magnitudes. They suggest that yellow cards increase by around 10% as a result of the incentive change, although in this case our estimates are somewhat imprecise.

Overall, we take these results as indicating that teams unambiguously increased the amount of sabotage.

## Net Effects of Increasing Attacking Play and Negative Play on Outcomes

We have found that because of the incentive change, whereas offensive effort increases, so does sabotage. In principle, it is not clear whether

**Table 8.3.** Changes in Sabotage Measures

| | Defenders | | | Fouls | | | Yellow Cards | | |
|---|---|---|---|---|---|---|---|---|---|
| Explanatory Variable | (I) | (II) | (III) | (IV) | (V) | (VI) | (VII) | (VIII) | (IX) |
| Incentive Change | 0.169* | 0.105 | 0.249*** | 1.99** | 2.1* | 2.2** | 0.23 | 0.421 | 0.217 |
| | (0.095) | (0.17) | (0.0965) | (0.93) | (1.2) | (0.989) | (0.282) | (0.459) | (0.276) |
| Cup Dummy | 0.23*** | 0.285** | 0.274*** | 1.17 | 1.42 | 1.48 | 0.33 | 0.741** | 0.462** |
| | (0.069) | (0.13) | (0.0713) | (0.856) | (1.13) | (0.908) | (0.202) | (0.294) | (0.191) |
| Year Effect | -0.291*** | -0.171 | -0.205** | -0.707 | -0.363 | -0.462 | 0.107 | -0.159 | 0.125 |
| | (0.0855) | (0.168) | (0.0913) | (0.897) | (1.18) | (0.96) | (0.27) | (0.455) | (0.267) |
| Visitor Dummy | 0.0725* | 0.0793** | 0.166 | 0.168 | | 0.411*** | 0.405*** | | |
| | (0.0423) | (0.0329) | (0.211) | (0.209) | | (0.069) | (0.0662) | | |
| Own Budget (000s) | -0.0363*** | | -0.0545 | | | -0.0124 | | | |
| | (0.00682) | | (0.0376) | | | (0.0135) | | | |
| Opponent's Budget (000s) | 0.016** | | -0.108*** | | | 0.0399*** | | | |
| | (0.00706) | | (0.0369) | | | (0.0135) | | | |
| Intercept | 4.05*** | 4.05*** | 3.93*** | 16.2*** | 16.4*** | 15.9*** | 2.33*** | 2.07*** | 2.11*** |
| | (0.0286) | (0.0388) | (0.0324) | (0.191) | (0.236) | (0.229) | (0.0549) | (0.0768) | (0.0671) |
| Include team fixed effects? | No | No | Yes | No | No | Yes | No | No | Yes |
| *N* | 1698 | 1574 | 1698 | 1596 | 1568 | 1596 | 1716 | 1572 | 1716 |
| $R^2$ | 0.012 | 0.031 | 0.352 | 0.020 | 0.026 | 0.046 | 0.013 | 0.042 | 0.077 |

*Note*: This table reports differences-in-differences estimates of the effect of the change in incentives on the number of defenders initially deployed by a team, the number of fouls committed by the team, and the number of yellow cards received by the team. The unit of observation is a team within a match. The first difference compares matches in seasons before and after the rule change, and the second difference compares matches in the cup tournament to league play. Standard errors clustered on matches are reported in parentheses. * denotes significant at the 10% level, ** at the 5% level, and *** at the 1% level.

these changes may lead to more goals, fewer goals, or to no change in the number of goals. Interestingly, we find in columns I–III in table 8.4 that there is no significant change in the number of goals after the change in incentives in any of the specifications. Hence, the increase in attacking play was not enough, given the increase in sabotage, to increase goals. The effect is quite precisely estimated at around zero.

Columns IV–XII in this table present results for some other outcome measures of interest:

1. The proportion of ties did not decrease, even though such a decrease would be ex ante Pareto preferred by all teams.[9]
2. Extra time, which is awarded at the discretion of the referees to compensate for interruptions in play, does *increase* as a result of the incentive change. Because most interruptions are caused by fouls and yellow cards, especially those that cause injuries, this increase is further, indirect evidence of sabotage.
3. Finally, there still is the question of whether the public preferred the increase in more physical play. Attendance measures this margin. Our findings suggest that the incentive change actually *decreased* attendance to the stadium. Note that the most complete specification, which controls for the popularity of the teams using a full set of home and visiting team fixed effects, is the one that gives the clearest result. We will return later to this issue and examine which actions may have led to lower attendance, that is, to reducing welfare as perceived by FIFA. We first try to get a better understanding of why goals did not change after the change in incentives by investigating the dynamic strategic mechanism underlying the changes in behavior we have documented.

## COMPETITION DYNAMICS: HOW DID SABOTAGE KEEP GOALS FROM INCREASING?

We study here the dynamics of the competition using the variables for which there exists information on their timing during the match: player substitutions and goals.

### Player Substitutions during the Game

Figures 8.2A and 8.2B present graphically the DID estimates of the changes in the number of defenders and attackers by game score. Although any player can defend and attack, changes in strategies

9 Increasing attackers and defenders, therefore, does not increase the risk of the outcome, except for the case of scoreless ties (not shown), which do decrease.

**Table 8.4.** Net Effects of Incentive Change on Goals and Other Outcome Variables

| | Goals Scored | | | Tie Indicator | | | Extra Time | | | Attendance | | |
|---|---|---|---|---|---|---|---|---|---|---|---|---|
| Explanatory Variable | (I) | (II) | (III) | (IV) | (V) | (VI) | (VII) | (VIII) | (IX) | (X) | (XI) | (XII) |
| Incentive Change | −0.0202 | 0.00187 | 0.011 | 0.0435 | −0.0534 | 0.00213 | 0.471** | 0.491** | 0.503* | −0.0104 | −0.0711* | −0.103*** |
| | (0.169) | (0.272) | (0.177) | (0.0859) | (0.143) | (0.0952) | (0.234) | (0.241) | (0.259) | (0.0279) | (0.0373) | (0.0367) |
| Cup Dummy | 6.629E-4 | −0.126 | −0.0169 | −0.0331 | −0.0504 | −0.0402 | −0.382** | −0.419** | −0.436** | 0.195*** | 0.148*** | 0.0837* |
| | (0.11) | (0.149) | (0.119) | (0.0652) | (0.104) | (0.073) | (0.163) | (0.177) | (0.201) | (0.0182) | (0.026) | (0.0429) |
| Year Effect | 0.0846 | −0.154 | 0.108 | −0.0856 | 0.0635 | −0.0088 | 0.0354 | 0.0077 | −0.0378 | −0.025 | 0.0397 | 0.0386 |
| | (0.158) | (0.266) | (0.17) | (0.0795) | (0.14) | (0.0898) | (0.217) | (0.226) | (0.248) | (0.0249) | (0.0368) | (0.0467) |
| Intercept | 1.25*** | 1.37*** | 1.5*** | 0.297*** | 0.334*** | 1.71*** | 3.46*** | 3.4*** | 4.37*** | 0.755*** | 0.362*** | 1.24*** |
| | (0.0443) | (0.063) | (0.0598) | (0.0238) | (0.0282) | (0.21) | (0.0648) | (0.127) | (0.204) | (0.00846) | (0.0636) | (0.0579) |
| Additional Controls | | Home and visitor budgets, visitor dummy | Visitor dummy | | Home and visitor budgets | | | | Stadium capacity, home and visitor budgets | | Home and visitor goals, yellow cards, and red cards | Home and visitor goals, yellow cards, and red cards |
| Additional Fixed Effects | | | Team | | | Home and visiting team | | Home and visiting team | Home and visiting team | | | Home team |
| $N$ | 1718 | 1574 | 1718 | 859 | 787 | 859 | 801 | 800 | 800 | 801 | 801 | 787 |
| $R^2$ | 0.001 | 0.102 | 0.125 | 0.005 | 0.010 | 0.074 | 0.057 | 0.085 | 0.121 | 0.085 | 0.689 | 0.692 |

*Note*: This table reports differences-in-differences estimates of the effect of the incentive change on the number of goals scored by a team, the probability of a tie match, the number of extra minutes added to the match by referees, and match attendance. Attendance is measured as the proportion of the available seats in the stadium that were occupied. The first difference compares matches in seasons before and after the rule change, and the second difference compares matches in the cup tournament to league play. Standard errors clustered on matches are reported in parentheses. The unit of observation is the team in the match in the first three columns and the match in the remaining columns. * denotes significant at the 10% level, ** at the 5% level, and *** at the 1% level. For columns IV–VI, estimation using a probit model generates comparable results.

during the game are better implemented by substituting in new specialists. Using the evidence on player substitutions during the game, we find that the number of defenders used by a team in the lead increases monotonically with the size of the goal difference. Conversely, teams use more attackers the further behind they fall, and this relationship is also monotonic.

Figure 8.2A, which shows the effect of the change in the number of defenders by goal score (where the number is measured relative to the

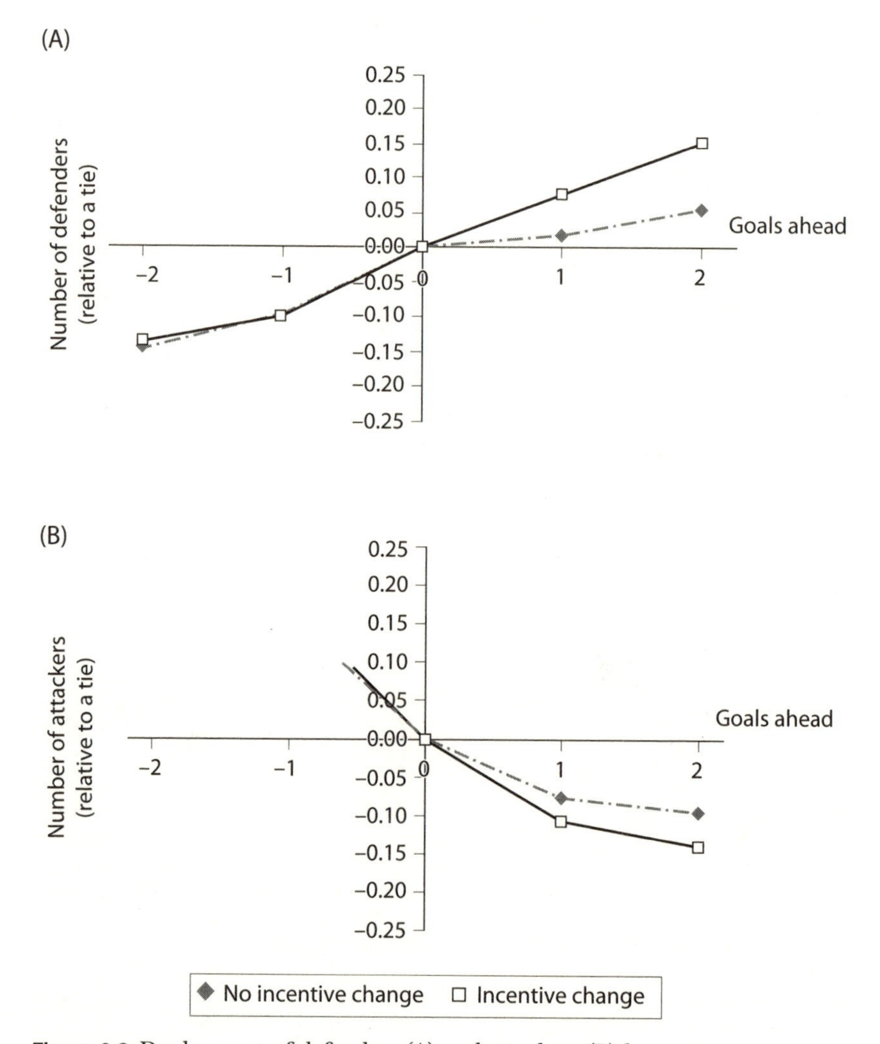


**Figure 8.2.** Deployment of defenders (A) and attackers (B) by game score.

number used in a tie), clarifies how teams are adapting their strategy to the new situation. After the incentive change, teams that get ahead in the score by one or two goals increase significantly their deployment of defenders relative to such deployment before the rule change. For 1 goal ahead, the test statistic for the equality of the number of defenders is $F(1, 858) = 5.64$, and $p$-value is 0.017, whereas for 2 goals ahead, it is $F(1, 858) = 4.26$, and $p$-value is 0.039. That is, when a team is ahead, it deploys a strategy that aims at conserving the score relative to the possibility of scoring more goals.

Moreover, recall that teams were already using more defenders in the initial lineup. Hence, the change relative to the old reward scheme is even more significant.

Figure 8.2B shows the change in the deployment of the number of attackers by game score, again relative to the number used in a tie. The change goes in the same direction of more conservatism when ahead, and it has a similar size.

After the incentive change, a team deploys 0.1 fewer attackers when it is ahead than when it is tied, although the drop is not statistically significant (for 1 goal, the $p$-value is 0.310, and for 2 goals it is 0.416).[10]

## Likelihood of Scoring and Goal Attempts during the Game

Figures 8.3A and 8.3B report the estimated coefficients of two different regressions of goals and shots aimed at the opponent's goal.

Figure 8.3A presents the DID estimates of the probability of scoring by game score. Consistent with its increasing defensive stance, the team ahead was less likely to score a goal after the rule change. This change is statistically significant (for 1 goal ahead, the test on the equality of the scoring probability is $\chi^2(1) = 5.46$, and $p$-value is 0.019; for 2 goals ahead,

10 These figures report the estimated coefficients from a regression of the number of defenders (A) and attackers (B) on an indicator variable for the incentive change interacted with indicators for the number of goals ahead or behind as well as team, minute, year, cup game, and match fixed effects. The unit of observation is one minute of play by a team in a match. The regressions contain 154,620 observations with an $R^2$ of 0.226 (for A) and 0.228 (for B). The reported coefficients are relative to the number of defenders (A) or attackers (B) used during a tie. For instance, the point (1, 0.077) on "Incentive Change" in figure 8.2A means that after the change, teams on average had 0.077 more attackers on the field during minutes when they were ahead in the score than during minutes when the game was tied. Similarly, for figure 8.2B, when teams are 1 or 2 goals behind, $F$-tests using standard errors clustered on match fail to reject the equality of coefficients pre- and post-rule change in either figure. When teams are 1 or 2 goals ahead, the pre- and post-rule change coefficients are statistically different at the 0.05% level in figure 8.2A but not statistically different in figure 8.2B.

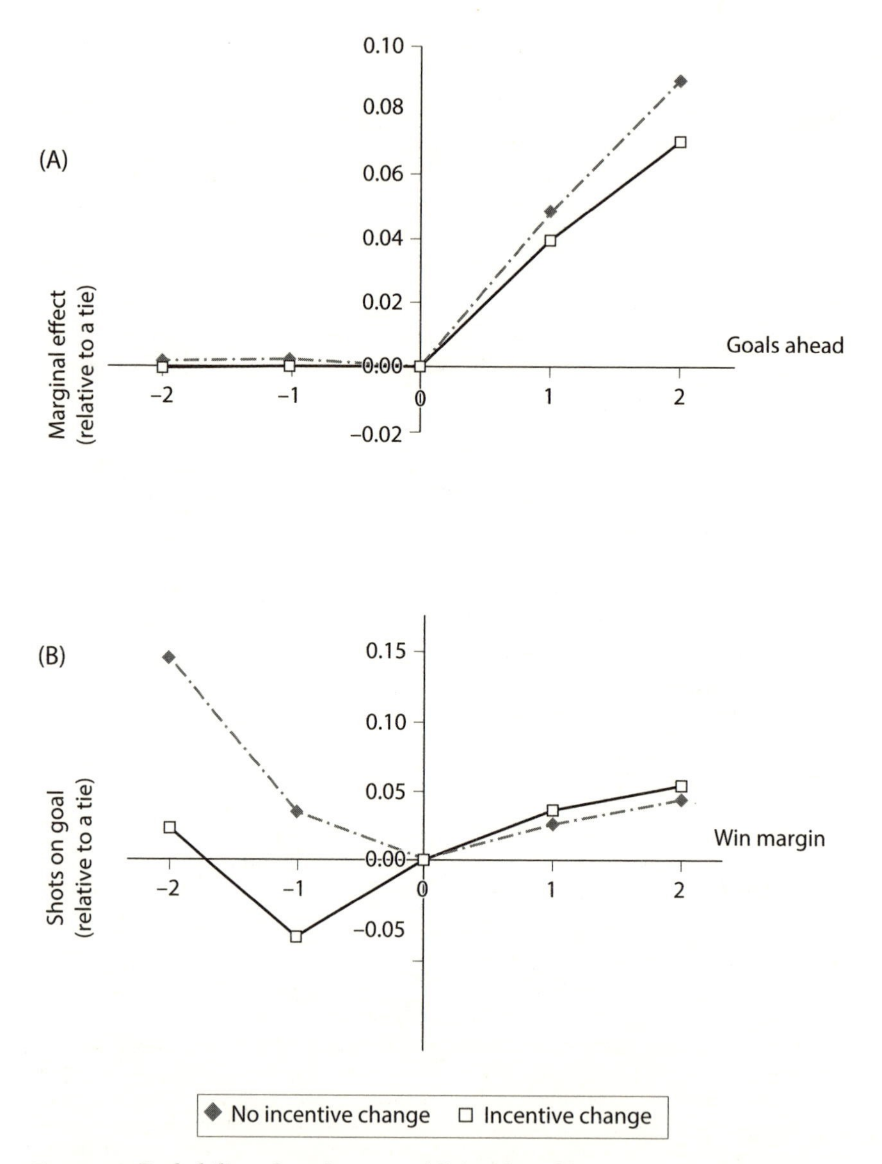


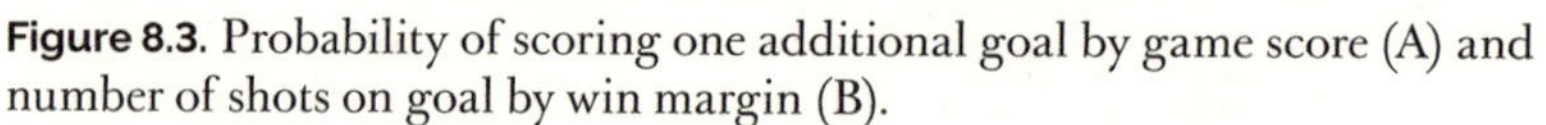


**Figure 8.3.** Probability of scoring one additional goal by game score (A) and number of shots on goal by win margin (B).

$\chi^2(1) = 4.09$, and *p*-value is 0.043). Since the probability that the team behind scores a goal in any particular minute is very small, the team that is behind suffers only a tiny decrease in the probability of scoring as a result of the increasingly aggressive defensive stance of the team ahead. Yet, the change is transparent.[11]

Figure 8.3B presents additional, indirect evidence of this drop. Because no records exist of shots per minute in the data set, the figure shows the number of shots over the entire match. The behavior is U-shaped: Teams take more shots both in matches where they end up behind and in matches where they end up ahead. After the incentive change, the total number of shots taken by a team that ends up losing decreases significantly (for 2 goals, $F(1, 797) = 5.42$, and *p*-value is 0.020; for 1 goal, $F(1, 797) = 7.51$, and *p*-value is 0.006), and there is no change for the team that ends up winning. Although, of course, the match outcome is endogenous to the number of shots, we find that this evidence complements that in the previous figure.

To summarize, teams ahead use fewer forwards and more defenders after the incentive change, score fewer goals, and allow, overall, a smaller number of shots by their opponents. Does this change contribute to making the beautiful game more or less beautiful?

11 Figure A reports the estimated coefficients from probit regressions of an indicator equal to 1 in minutes in which a team scored on an indicator variable for the incentive change interacted with indicators for the number of goals ahead or behind, as well as team, minute, year, and cup game fixed effects. The unit of observation is one minute of play by a team in a match. The regression contains 153,959 observations. The probit coefficients have been transformed to marginal effects at the mean of each indicator and are reported relative to ties. The point (1, 0.048) on "No Incentive Change," for example, means that before the incentive change, teams on average were 4.8% more likely to score a goal during minutes when they were ahead than during minutes when the game was tied. When teams are 1 or 2 goals behind, *F*-tests using standard errors clustered on match fail to reject the equality of coefficients pre- and post-rule change. When teams are 1 or 2 goals ahead, the pre- and post-rule change coefficients are statistically different at the 0.05% level. Figure B reports the estimated coefficients from regressions of the number of shots on goal on an indicator variable for the incentive change interacted with indicators for the margin of victory as well as team, year, win margin, and cup game fixed effects. The unit of observation is a team in a match. The regression contains 1,596 observations with an $R^2$ of 0.108. The reported coefficients are relative to the number of shots on goal made in games that were tied. For win margins of 1 and 2 goals, *F*-tests using standard errors clustered on match fail to reject the equality of coefficients pre- and post-rule change. For loss margins of 1 and 2 goals, the equality of coefficients can be rejected at the 5% level.

## DYSFUNCTIONAL RESPONSE OR DESIRABLE INTENSITY?

It seems reasonable to conclude from the evidence that as a result of the incentive change, effort increased, teams engaged in a more intense and physical type of play, and more "dirty" actions took place. Yet sabotage activities need not be detrimental to the game. That is, it is unclear whether or not this behavior by the agents is "bad" from the perspective of the principal. Contrary to the provision of incentives in firms and other organizations, where *any* amount of sabotage is undesirable for the principal, in a sports context some strong physical play may be desirable. For instance, it is often argued that physical play and brawls are desired by the public in ice hockey. This, despite FIFA's stated purpose for the incentive change, could also be the case in soccer.

Here we study the extent to which the public dislikes the increase in dirty play after the incentive change. To do this, we exploit a useful feature of league play: All teams are allocated to all stadiums, until they each play in every other team's home stadium. This feature allows us to tease out the effect of playing against a dirtier rival—that is, one that undertakes more sabotage actions—on attendance at the stadium and on TV audiences.

Table 8.5 studies the effect of playing against a dirtier team at one's home stadium, that is, the response of fans attending at the stadium to the expected "dirtiness" of the visiting team. We proxy for this using the average number of fouls, yellow cards, and red cards by the visitor during the season in question. We also compute an index of sabotage propensity by a team using a factor analysis on the matrix of these three variables and picking the first principal component. Table 8.6 reports the net effect of the incentive change on TV audiences and, using the same principal component, also the effect of dirtier games on these audiences.

The results show that both stadium and TV audiences declined as a result of dirtier play, even after controlling for the losing or winning record of the teams in the match and other variables.[12] These findings, together with the result in table 8.4 showing how stadium attendance declined as a result of the incentive change, allow us to conclude that stronger incentives to win led to dirtier play, which turned off stadium

12 The results are strong for every variable except for red cards, which exhibit high standard errors. Red cards, however, represent a small proportion of all sabotage activities (less than 1%) and are to a large extent random and unplanned, in that they involve unusual behavior (e.g., insulting, spitting) that is clearly beyond the bounds of the game. We have also put budget control variables in regressions X, XI, and XII of table 8.4 and in the regressions of table 8.5 and found no significant differences from the results without these controls.

**Table 8.5.** Stadium Attendance and Sabotage

| Explanatory Variable | (I) | (II) | (III) | (IV) | (V) | (VI) | (VII) | (VIII) |
|---|---|---|---|---|---|---|---|---|
| Measures of Visitor's Dirty Play | | | | | | | | |
| • Fouls | −0.00836** | −0.00693* | | | | | | |
| | (0.00424) | (0.00369) | | | | | | |
| • Yellow Cards | | | −0.0191* | −0.0202* | | | | |
| | | | (0.0108) | (0.0103) | | | | |
| • Red Cards | | | | | −0.0484 | 0.0949 | | |
| | | | | | (0.0934) | (0.0916) | | |
| • Dirtiness Index | | | | | | | −0.0435*** | −0.0345** |
| | | | | | | | (0.0158) | (0.0141) |
| Home Team Wins | | 0.00243 | | 0.00213 | | 0.00219 | | 0.00235 |
| | | (0.00241) | | (0.00241) | | (0.00241) | | (0.00242) |
| Visitor Wins | | 0.0149*** | | 0.0151*** | | 0.0153*** | | 0.0148*** |
| | | (0.00157) | | (0.00157) | | (0.00166) | | (0.00155) |
| Season Indicator | | −0.00893 | | −0.00913 | | −0.0132 | | −0.00597 |
| | | (0.0115) | | (0.0117) | | (0.0114) | | (0.0119) |
| Stadium Capacity | | −0.0121*** | | −0.0121*** | | −0.0121*** | | −0.0121*** |
| | | (0.0037) | | (0.00369) | | (0.0037) | | (0.00368) |
| Game Number | | −0.00592*** | | −0.00588*** | | −0.00598*** | | −0.00585*** |
| | | (0.00103) | | (0.00103) | | (0.00105) | | (0.00102) |
| Intercept | 0.879*** | 1.32*** | 0.789*** | 1.26*** | 0.741*** | 1.2*** | 0.737*** | 1.2*** |
| | (0.0724) | (0.159) | (0.0306) | (0.143) | (0.00853) | (0.141) | (0.00458) | (0.141) |
| $N$ | 750 | 750 | 750 | 750 | 750 | 750 | 750 | 750 |
| $R^2$ | 0.503 | 0.565 | 0.501 | 0.565 | 0.499 | 0.563 | 0.505 | 0.566 |

*Note*: This table reports regression estimates of the effect of the visiting team's dirty play on attendance. The unit of observation is a match, and the sample is restricted to league matches. Each of the measures of the visitor's dirty play is constructed as averages over the season in which the match took place. The "Dirtiness Index" is the first principal component of fouls, yellow cards, and red cards. "Home Team Wins" and "Visitor Wins" are the number of wins by each team in the match within the same season before the game in question. "Stadium Capacity" is measured in number of seats. "Game Number" is the game number in the season. All specifications include home-team fixed effects. * denotes significant at the 10% level, ** at the 5% level, and *** at the 1% level.

**Table 8.6.** TV Audience (Share) and Sabotage

| Panel A | | |
|---|---|---|
| Constant | 0.880*** | 1.251*** |
| | (0.127) | (0.111) |
| Incentive Change | -0.202*** | -0.217*** |
| | (0.057) | (0.040) |
| Cup Dummy | 0.152*** | 0.097** |
| | (0.045) | (0.040) |
| $R^2$ | 0.532 | 0.571 |
| $N$ | 801 | 707 |

| Panel B | | |
|---|---|---|
| Constant | 0.828*** | 1.332*** |
| | (0.112) | (0.234) |
| Dirtiness Index | -0.062*** | -0.079*** |
| | (0.019) | (0.023) |
| Home Team Wins | | 0.003 |
| | | (0.165) |
| Visiting Team Wins | | 0.014*** |
| | | (0.003) |
| Season Indicator | | -0.001 |
| | | (0.002) |
| Stadium Capacity | | -0.012** |
| | | (0.004) |
| Game Number | | -0.000 |
| | | (0.002) |
| $R^2$ | 0.352 | 0.397 |
| $N$ | 297 | 297 |

*Notes*: This table reports regression estimates of the effect of the incentive change (panel A) and dirty play (panel B) on TV share. Panel A includes controls for home and visitor goals and yellow and red cards in the first column, and in the second, in addition, home team fixed effects. In panel B, the "Dirtiness Index" is the first principal component of fouls, yellow cards, and red cards for both teams. "Home Team Wins" and "Visiting Team Wins" are the number of wins by each team in the same season up to the game in question. "Stadium Capacity" is measured in number of seats. "Game Number" is the game number in the season. All specifications include home-team fixed effects. Standard errors are clustered. ** denotes significant at the 5% level, and *** at the 1% level.

attendances and TV audiences. As such, these strong incentives did have dysfunctional consequences. Rough back-of-the-envelope calculations suggest, for instance, that the amount of fans who would not go to the stadium at the average league match because of the estimated increase in "dirtiness" induced by the three-point rule was about 6% to 8%. Similarly, the decrease in TV audiences as a result of the incentive change can be estimated to be in the range of 2% to 4% on average. There is no easy way to gauge the overall economic effect, but roughly, these findings suggest that the league may have become around 5% poorer as a result of the incentive change. Of course, this estimate does not take into account the subsequent cumulative effects that this change may have had in future seasons. No doubt this is an important question for future research.

*

Although traditionally most of the literature on incentives emphasized the trade-off between risks and incentives, empirical evidence for the importance of such trade-off is tenuous (Prendergast 2002). The more modern view (Lazear 1989; Holmstrom and Milgrom 1991, 1994; and Baker 1992) emphasizes the limits placed on the strength of incentives by the difficulty in measuring output correctly and the costs that may be incurred when, as a reaction to stronger incentives, agents reoptimize away from the principal's objective.

We see this chapter as providing a strong empirical endorsement for this view. We find that an increase in the reward for winning increased, counter to FIFA's intentions, the amount of sabotage effort undertaken by teams. Although there appears to be some increase in attacking effort, no actual change took place in the variable where change was intended, goals scored. The mechanism underlying these patterns is increasing conservatism: Teams try to preserve their lead by freezing the game. The decrease in stadium attendance and TV audience we find means that stronger incentives turn out to be *detrimental* to the game.

Although theoretical research warns about the possible harmful effects of increasing incentives when workers can engage in sabotage, the theory has remained untested in the literature until the results in this chapter were presented at various academic conferences and seminars (see Chowdhury and Gürtler 2013 for an excellent survey). Workers may indeed bad-mouth their colleagues and actively prevent them from achieving good results by withholding information and other means. However, they typically do their best to conceal their efforts. For this reason, evidence on sabotage activities is, by its nature, at best anecdotal. In the natural setting we have studied, however, both productive and destructive actions can be observed. Moreover, a critical advantage is that we can study the effects of a change in incentives using a control group to eliminate any effects unrelated to the incentive change.

Viewed from this perspective, the analysis may be interpreted as providing the first explicit empirical test of worker-incentive problems in a natural multitask setting, where tasks can be productive and destructive.

It is unclear, however, whether the evidence can be interpreted as a test of the standard tournament model with both productive and destructive actions, as in Lazear (1989). The purpose of the rules and their enforcement in professional sports is to make that single-minded pursuit of winning entertaining for the viewers. The tournament, in this sense, more than the compensation mechanism, is simply the product.[13] Put differently, using sporting rules to reward teams based on their level of entertainment rather than on their winning may in fact defeat the purpose of the product.

Lastly, these findings are useful to make an educated guess as to how teams would respond to future potential changes in rules. For instance, take the proposal at the English Premier League annual meeting in 2005 that 4 points be awarded for *away* wins, rather than 3. The mechanisms for potential unintended consequences that our analyses have uncovered suggest that this is probably a bad idea, and one should proceed with caution. Overall, to the extent that rule changes have been discussed and decided on with very little data and even less analysis, the study in this chapter represents a contribution to the discussion.

*

Major forms of sabotage activities are often illegal and hard, if not impossible, to document even in the setting we have studied. Two anecdotes from World Cup games testify to the difficulties of obtaining such data.

The first one comes from Relaño (2010) and involves two of the greatest players ever. One of the many things that Diego Armando Maradona did after retiring from football was a quite successful TV program, *La Noche del 10*, in Argentina. To give prominence to the first show, he invited Pelé for an interview. The two players never had a great relationship (they still do not), always jealous of each other and disputing who was better than whom in soccer history. Pelé was paid 48,000 euros for attending the program (Maradona charged 40,000 per appearance, and so Pelé demanded 20% more).

13 Tournaments where workers can allocate their time and attention only in the direction of productive activities were introduced by Lazear and Rosen (1981). See also Green and Stokey (1983), Rosen (1986), and Prendergast (1999) for a review. For empirical work on tournaments in a sports context with only productive activities, see Ehrenberg and Bognanno (1990), and for experimental work, see Bull et al. (1987). Theoretical work with multiple productive activities, such as individual and cooperative tasks, appears in Itoh (1991, 1992), and Rob and Zemsky (2002). Drago and Garvey (1998) use survey data to study helping others on the job.

The interview started. After a lengthy exchange of compliments and courtesies that sounded pretty fake, Pelé suddenly changed the game. "I have a question, and I hope you will be honest with me: Did you put sleeping pills in the water bottle for Branco?" (Pelé referred to a known issue. In the Argentina–Brazil game of the World Cup in Italy on June 24, 1990, when the Argentine masseur Galíndez went off the bench to assist his player Troglio, he also used the opportunity to give intoxicated water to Branco, the Brazil captain. Galíndez had bottles of water with two types of caps: Blue (good water) and yellow (water with sleeping pills), which should be given to the Brazilians if they asked for water. And so it happened. Branco asked for water and was given a bottle with a yellow cap. He felt bad during the rest of the game. Branco then later recalled that he was somewhat surprised when an Argentinian player took a bottle of water and a teammate told him, "No, not from that one." A few months before that program, on another TV program (*Mar de Fondo*), Maradona confirmed the suspicions: "Someone put Rohypnol in the water and everything came apart. . . . Branco did not greet me any more after drinking from that bottle.")

Coming back to the interview, Maradona was visibly taken aback by Pelé's question. Either he denied the accusation and was a liar, or he recognized a major form of sabotage from Argentina to Brazil. "I was not there . . . yes, something happened. . . ." Pelé insisted, and Maradona kept dribbling. "We acknowledge the sin but do not report the sinner. . . ." Until he found a sentence that allowed him to escape on top: "I never had to put to sleep anyone to win a game." It broke the biggest applause of the night. Argentina won the match with a lone goal by Claudio Caniggia in a famous play in which the entire Brazilian defense went chasing Maradona, who took the opportunity to pass to his teammate entirely unmarked.

The second one involves what is probably the biggest shock in the history of the World Cup: In 1966, North Korea beat Italy, eliminating it from that year's World Cup tournament in England.

Italy was one of soccer's most successful teams since winning back to back World Cups in the 1930s. Little was known about the Asian team before the tournament, but few expected them to provide much opposition to an Italian side featuring A.C. Milan star Gianni Rivera (a future European player of the year), Sandro Mazzola (son of Valentino Mazzola, the former Italian team captain) and Giacinto Facchetti (the F.C. Internationale Milano, or Inter Milan, icon). On July 19, 1966, however, in front of 18,000 spectators crammed into Middlesbrough's Ayresome Park, Italy lost to World Cup debutants North Korea. This was the first time that a nation from outside Europe or the Americas had progressed from the first stage of a World Cup to the next round. The reasons for this shock remain unclear. No substitutes were allowed in the tournament, so when Giacomo Bulgarelli was stretchered off after 30 minutes after a knee injury, the Italian team had to manage with just 10 men for more than an hour of play. One would think that this fact might have played an important role. But, interestingly, it was never even mentioned as an excuse for the defeat. Instead, back in Italy some players reported to the media that they suspected (but could not prove) the most creative form of sabotage: It seemed to them that at halftime North Korea had replaced all 11 of their players in the lineup!

# 9

# FEAR PITCH

© TAWNG/STOCKFRESH

If we let things terrify us, life will not be worth living.
—Lucius Annaeus Seneca, *Letters from a Stoic*

Think firecrackers, flying bottles, flares raging through the stands, agitated police dogs, and the lingering threat of violence. Think homemade bombs, street battles, and prearranged match-day fights with stones, baseball bats, and knives. Think racist chanting and swastikas.

Chances are you will find them at these games: Flamengo vs. Fluminense in Brazil, Ajax vs. Feyenoord in Holland, Olympiakos vs. Panathinaikos in Greece, Liverpool vs. Manchester United in England, Red Star vs. Partizan in Serbia, Barcelona vs. Real Madrid in Spain, Celtic vs. Rangers in Scotland, Galatasaray vs. Fenerbahçe in Turkey, Lazio vs. Roma in Italy, and Boca Juniors vs. River Plate in Argentina. According to a sport magazine recently, these are the top 10 most venom-filled soccer enmities in the world.

One could easily think of the top 100 or find a hundred examples when deaths and arrests followed the sheer terror that violence generated. Some extreme cases immediately come to mind, such as the Heysel Stadium disaster on May 29, 1995, where a major brawl erupted before the European Cup Final between Liverpool F.C. (England) and Juventus F.C. (Italy), in which 39 supporters died and 600 were injured. Tragic cases with fatal casualties abound in the history of soccer in all countries, and rare is the country with no deaths even in just the past few years. Cases with no deaths, just arrests, fights, injured fans, and basic vandalism are even more frequent year after year in every country. Take, for instance, the case of an Inter Milan fan, Matteo Saronni, from the village of Isola di Fondra. Matteo was arrested for throwing a motorbike from the upper stands at the San Siro Stadium in 2001 to the lower stands. Surprisingly, no one died. This fact is probably why he escaped punishment. A judge allowed him off with a warning while the club banned him for a year. What did he have to do to get some time in jail? Perhaps throw something larger than a motorbike? Would a Honda Civic have been sufficient? Two years later, he was arrested for throwing a missile at a match against Juventus, and four years later for throwing a flare at an abandoned Champions League match against A.C. Milan. In both cases, no one died.

Many countries in the world are plagued by hooligan violence and, logically, potential spectators avoid attending games in fear of this violence (Priks 2008). Fear, often considered the oldest and strongest emotion of humankind, is the topic of this chapter.

How rational is fear? Is it purely emotional? Is it feasible to conquer fear? If so, how and when is it done? At what price? Who is more likely to overcome fear? When? These are too many questions, with difficult answers and, what is worse, with little or no theoretical modeling that may be able to guide empirical work and interpret the data.

No modeling, that is, until Becker and Rubinstein (2013). These authors offer the first rational approach to the economics and psychology of fear by considering not just the effect of danger on distorting subjective beliefs away from objective probabilities of risk (as typically considered in psychology) but, more importantly, that people have the incentives, and can sometimes learn, to control fear. A key conceptual breakthrough in the model is that a person's willingness to "invest" in controlling fear depends on the economic costs and benefits associated with acquiring this self-control. Thus, because in the Becker–Rubinstein model the incentives associated with controlling fear differ across individuals in predictable ways, the model provides a number of testable implications.

Contrary to a number of chapters in this book, a soccer setting is not the first setting where the theory under study has been tested. It is a new setting but not the first one. Becker and Rubinstein already tested the main predictions empirically using the response of Israelis to terror incidents during the Al-Aqsa Intifada. Also known as the Second Intifada, this was the second Palestinian uprising—a period of intensified Palestinian–Israeli violence, which began in late September 2000 and ended around 2005. The death toll, including both military and civilian, is estimated to be about 3,000 Palestinians and 1,000 Israelis (Jews and Arabs), as well as 64 foreigners. Consistent with the theoretical predictions, Becker and Rubinstein find, for instance, that the overall effect of attacks on the usage of goods and services subject to terror attacks (buses, malls, restaurants) reflects solely the reactions of *occasional* users and consumers. Terrorist attacks do *not* have any effect on the demand for these goods and services by *frequent* users and consumers. The reason is that frequent users are those who also tend to receive *greater* benefits from learning to overcome fear. Furthermore, once an individual learns to control fear triggered by, say, bus attacks, this control reduces the degree to which other types of terrorism (e.g., attacks in malls, coffee shops, or restaurants) cause her or his subjective and objective beliefs to diverge.

Fear, one of the small set of basic and innate emotions, is induced by a perceived threat and often has huge and enduring effects on human behavior. There is no question that fear is an emotion difficult to model. In fact, in economics neither the standard expected utility (EU) model of decision under risk, nor its state-dependent version, may explain why negligible changes in the probability of being harmed may have the arguably substantial effects that we often observe on individual choices. As a result, economists are often tempted to resort to behavioral explanations, including bounded rationality or even "irrationality," to try to account for the seemingly disproportionate response of people to, for example, mad cow disease, terrorist acts, or the H5N1 bird flu virus.

Becker and Rubinstein (2013) argue that the EU model should not be abandoned, it should be modified. The starting point in their analysis is that people are human: Emotions shape beliefs and behavior, so that subjective and objective beliefs can diverge. However, individuals can adjust too. Though it is costly and imperfect, people can learn to control their emotions, and they are more likely to do that when it is in their long-term interests. Thus the willingness to control one's emotions depends on the economic costs and benefits associated with acquiring this self-control. Even in a world with emotionally driven individuals, economic incentives contribute to shaping the degree to which emotions distort choices. When there are powerful incentives associated with learning to control one's emotions, a person's subjective beliefs about danger should be expected to get closer to the objective risks of those dangers.

A large body of evidence in psychology suggests that the capacity to control fear may be gained through training, past experience, and other forms of investment in this specific type of human capital. Because building one's capacity to deal with fear is costly and does not pay back the same returns to everyone, people differ in how much they invest in controlling fear. For example, frequent users of bus or airline services that are subject to terrorist attacks receive greater benefits from overcoming fear than occasional users. Therefore, frequent users may be more interested in investing in controlling fear and thus keep their consumption closely aligned with that indicated by the objective dangers. Occasional users, on the other hand, may be more interested in opting out of the "terror-infected good."

There is a literature in economics on "anticipatory feelings," in which subjects intentionally distort their beliefs because they derive an intrinsic benefit (or cost) from expecting a good (or bad) outcome (Caplin and Leahy 2001, 2004; and Caplin 2003). There is also a literature in economics and psychology on the concept of "ability" as an important input in the capacity to adjust to changes. This literature suggests that individuals with greater cognitive ability should be more likely to overcome fear because they are more likely to form subjective beliefs that are closer to objective probabilities, that is, less likely to "overreact." As we will see, incorporating this view into a rational model provides interesting testable implications. We first present the Becker–Rubinstein model.

Consider a situation where individuals consume two goods: a good ($x$) that is subject to fear caused by violence, and all other goods ($y$). Individuals live for one period. The probability of surviving to the end of the period is determined by their consumption plans. As long as they avoid consumption of $x$, their probability of survival equals 1. The more they consume of $x$, however, the less likely they are to survive.

Assume that fear exaggerates the subjective beliefs that the effect of consuming $x$ has on the probability of survival. People then respond to violence by reducing the consumption of $x$ and/or by taking costly actions to *control their fear.* Importantly, this "investment" is neither a "free lunch," nor does it pay the same return to every individual. Therefore, the optimal level of fear experienced by each individual is *endogenously* determined by the cost and benefit of controlling it.

Individuals' expected utility, $EU$, is defined as

$$EU = p(v,x,F)\ W(x,y)$$

where $p$ is the subjective probability of surviving to the end of the period, and $W$ is the utility from consumption of $x$ and $y$. The variable $p$ is adversely affected by the degree of violence, $v$, the consumption of the good subject to violence, $x$, and the emotion of fear, $F$. Formally, $p_v \leq 0$, $p_x \leq 0$, $p_F \leq 0$.

Furthermore, an increase in fear $F$ lowers the individual's belief that the effect of consuming $x$ has on the probability of survival: $p_{vx} \leq 0$. Violence, fear, and the consumption of $x$ are mutually reinforcing with respect to the effects on the subjective probability of survival: $p_{xF} \leq 0$, $p_{vF} \leq 0$.

For simplicity, $W$ is assumed to be a quasilinear function,

$$W(x,y) = a \cdot u(x) + y$$

where $u(x)$ is increasing and strictly concave, and the parameter $a$ is a shifter that captures the preference for $x$ relative to that for other goods $y$. The amount of fear $F$ is given by

$$F(v,k) = (1 - \alpha) \cdot x \cdot f(v,k)$$

where $k$ represents the knowledge of the violence activity and $\alpha$ is a binary variable that equals 1 if the individual chooses to control fear and 0 otherwise. Fear rises with the degree of violence $f_v > 0$, it is amplified by knowledge of violence, for instance, through media coverage, $f_k > 0$, and increases linearly with the consumption of $x$. In the absence of violence, there is no fear $f(0,k) = 0$.

Individuals can control their fears by spending a fixed amount of resources $\pi_a$. This amount is lower for individuals who possess greater abilities to assess objective risk accurately.

Therefore, the expected utility is

$$EU(x) = (1 - \alpha)\, EU^0(x) + \alpha\, EU^1(x)$$

where $EU^0(x)$ and $EU^1(x)$ represent the expected utility for $\alpha = 0$ and $\alpha = 1$, respectively. The budget constraint is

$$\pi_x x + y + \pi_\alpha \alpha = I$$

where $\pi_x$ is the price of $x$, $y$ is the numéraire (a basic standard by which value is computed), $\pi_a \alpha$ is the investment or expenditure to eliminate fear, and $I$ is income.

The expected marginal utility from consumption of $x$ is higher when investment is undertaken because the individual overcomes fear and brings his or her beliefs closer to objective probabilities:

$$dEU^1(x)/dx > dEU^0(x)/dx$$

This equation implies that the optimal consumption of $x$ when investing in reducing fear ($\alpha = 1$) is undertaken is always larger than the optimal consumption of $x$ without investing ($\alpha = 0$):

$$x^{1*} > x^{0*}$$

Two simple comparative statics readily arise.

1. The optimal consumption level always increases with the taste for $x$ (measured by $a$):

$$dx^{0*}/da > 0; dx^{1*}/da > 0$$

2. An increase in the degree of violence always reduces the consumption of $x$ by raising its implicit marginal cost:

$$dx^{0*}/dv < 0; dx^{1*}/dv < 0$$

The endogeneity of fear ($F$) is in the foundation of the Becker-Rubinstein model. Individuals choose consumption of $x$, $y$, and expenditure $\pi_a \alpha$ to maximize expected utility. They invest in controlling fear and overcome its distortive effect on their subjective beliefs if and only if

$$EU(x^{1*}) \geq EU(x^{0*})$$

Consider the following comparative statics that relate violence, consumption of $x$, and spending to reduce fear.

Consumers with greater taste for $x$ (as measured by $a$) are more likely to invest in overcoming fear. The reason is that an increase in the taste for $x$ induces an increase in $x$, which in turn raises fear, $F$. Therefore, this type of consumer benefits more from spending $\pi_a \alpha$ to reduce fear than those with a weaker taste for $x$. Foregone utility caused by distorted beliefs is also larger for those who benefit most from consuming $x$. Indeed, it can be shown that the expected benefits of investing in controlling fear, $EU(x^{1*}) - EU(x^{0*})$, are increasing in the taste for $x$:

$$d[EU(x^{1*}) - EU(x^{0*})]/da > 0$$

Let $\hat{a}$ denote the taste for $x$ for which consumers are indifferent between investing and not, all else being equal. Those with $a > \hat{a}$ spend $\pi_a$, overcome fear, and bring their beliefs closer to the objective danger. Those with $a < \hat{a}$ (all others) do not. As a result, the effect of an increase in the degree of violence on individuals' decisions to invest in controlling fear is ambiguous. When violence mainly increases objective dangers, individuals are less likely to invest in controlling fear. However, when violence has a negligible effect on objective danger but a large effect on subjective assessments of danger, then an increase in the degree of violence induces some individuals to invest in reducing fear.

The expected benefits of investing in controlling fear increase with the degree of violence if and only if the relative decline in the subjective probability to survive is larger than the relative rise in objective danger.

$$p_v^1/p^1 > p_v^2/p^2$$

Violence makes consumption of $x$ less attractive because it increases objective danger and intensifies fear for those who do not spend on controlling it. When violence increases objective dangers more than subjective dangers, consumers are less likely to invest in controlling fear and, therefore, reduce their consumption of $x$. When violence has a negligible effect on objective danger but a large effect on fear, then the effect of violence on consumption of $x$ is ambiguous for those who invest in controlling fear. As a result, an increase in the degree of violence reduces consumption of $x$ as long as it does not raise the expected benefit from investing in controlling fear. When violent incidents intensify fear, consumers with a *greater* taste for $x$ are *less likely* to change consumption plans, whereas those with a *smaller* taste for $x$ are more likely to overreact and substitute consumption of $x$ with consumption of all other goods ($y$).

Thus the model explicitly accounts for individuals' capacity to manage their emotions and control fear. Because managing emotions is costly and because the benefits of controlling fear differ across individuals, people differ in how much they invest in controlling fear in predictable ways. As indicated already, Becker and Rubinstein empirically evaluate the implications of their model by studying the reaction of Israelis to terror incidents during the Al-Aqsa Intifada. They differentiate between *frequent* users of goods and services that are subject to terrorist attacks and *occasional* users. Consistent with their model, it turns out that the overall effect of attacks on the use of services and goods subject to terror attacks is completely accounted for by the reactions of occasional users. Suicide attacks have no effect on the demand for these goods and services by frequent users.

Becker and Rubinstein also study the effect of education and media coverage on people's responses to terrorist attacks. If people with greater cognitive and noncognitive abilities have a greater capacity to assess risk and control fear, and if educational attainment is positively associated with cognitive and noncognitive skills, then more educated individuals are more likely to overcome fear and act as if they evaluate the risk associated with terror appropriately. This result is exactly what they find: The less educated are more likely to overreact to terrorist acts than more educated individuals.

How does knowledge of terror activities affect the public? Here they look at media coverage. Using the natural variation in the exposure of the Israeli population to newspapers (they are not printed on Saturdays or holidays), there is a large effect of suicide attacks during regular media coverage days but almost no effect when they are followed by either a holiday or a weekend. Moreover, the large effect of suicide attacks followed by regular weekdays' media coverage is found mainly in the use of bus services by the less educated families and among occasional users of bus services.

This chapter tests the same implications in a soccer setting. We compare the behavior of fans *before* and *after* games who were subject to acts of hooliganism, vandalism, or criminal damage, including those who experienced disturbances and violent confrontations among spectators.

The model predicts that

1. *Frequent* consumers of soccer matches should respond *less* to acts of violence (hooliganism, vandalism, and criminal damage) than *occasional* consumers of football matches.
2. Violence should have a differential effect across singles versus married individuals. It should *decrease* attendance by *married* fans (those with greater costs) but have a much smaller effect, perhaps even no effect, on the behavior of *single* individuals.
3. Acts of violence, hooliganism, and vandalism imply that the likelihood that a fan will renew his or her pass for the following season should be greater for singles than for married individuals. Similarly, singles should be more likely to become *new* season ticket holders than married individuals.
4. Fans with *greater* cognitive ability should respond *less* to acts of violence, hooliganism, and vandalism than those with lower cognitive ability.
5. Because of differences in media coverage, the response to acts of violence, hooliganism, and vandalism should be lower in games that end after midnight than in games that end before midnight.

We discuss and study these predictions next.

In soccer, as in many sports, there are season tickets and match-day tickets. A season ticket typically grants the holder access to all regular-season home games for one season without additional charges. It usually offers a discounted price over purchasing a match-day ticket for a given individual home game, and season ticket holders are usually allowed to buy tickets for other home matches (such as a cup tournament, or international tournaments or play-off games) earlier than other fans. They are also typically given priority when buying tickets for the away games.

We look at data from La Liga in Spain, where most soccer clubs during the period of analysis have three types of tickets: Socios, abonados, and single match-day tickets. *Socios* are season ticket holders who have to pay two fees: an annual fee each season and an initial lump-sum fee at the time they become socios, which is not reimbursed and allows the person to be a season ticket holder forever, if he or she wants to, by paying the corresponding annual fee each season. They have voting rights. *Abonados* are also season ticket holders but just for one season. They do not have to pay the lump-sum fee, which means that they are not guaranteed to be a season ticket holder the following season and do not have voting rights. Finally, there are just *single match-day tickets* that entitle the purchaser to attend one game. Socios are offered preferred seating in the stadium, then abonados, and then single ticket holders.

To become a socio or abonado, but not to buy a ticket for a single game, an individual has to fill out an application form that requests personal information (including education and employment) and enclose a copy of his or her national identity (ID) card. Until recently, the national ID card in Spain also included information on the marital status and the profession of the person at the time they got their ID card.

The data encompass 167 events during the period 1951–95 that were characterized by acts of violence, hooliganism, vandalism, and criminal damage, and for which the home teams could provide the required attendance data. For these events, we collect data for the home games before and after the game of the event, that is, 334 matches, as well as ticket holder data for the relevant seasons for each of the corresponding teams during those 45 seasons. The data are aggregated across all the events and converted into percentage terms to provide a simple visualization of the effects.

*Hypothesis 1. Frequent consumers of soccer matches should respond less to acts of violence (hooliganism, vandalism, and criminal damage) than occasional consumers of football matches.*

The data are consistent with this hypothesis (see figure 9.1). Attendance, on average, drops 3 percentage points for socios, 10 percentage points for abonados, and a sizable 40% for individuals who buy single match-day tickets. This basic finding remains unchanged if we control for a number of observed characteristics.

*Hypothesis 2. Violence should reduce the attendance by married fans but have a much smaller effect on the behavior of single individuals.*

The aggregate evidence also seems consistent with this hypothesis, although perhaps not as strongly (see figure 9.2).

Note, however, that socios can get divorced and/or married after they become socios. And so the data on the marital status at the

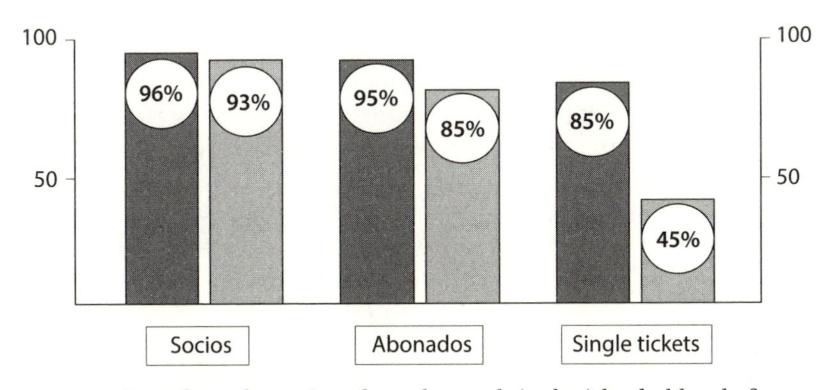


**Figure 9.1.** Attendance by socios, abonados, and single ticket holders before (dark) and after (light) acts of violence and hooliganism.

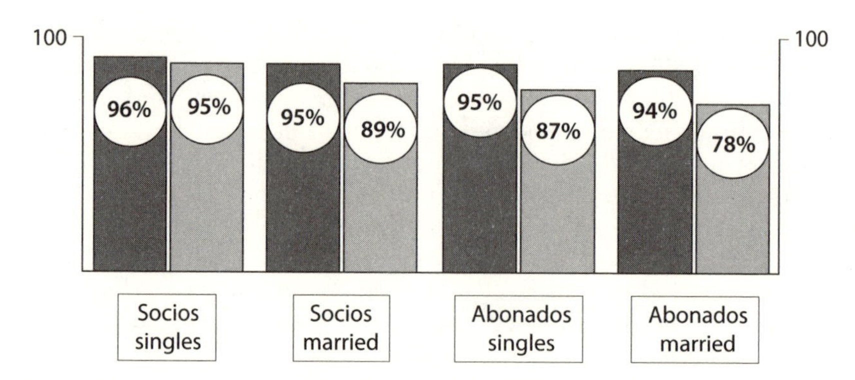


**Figure 9.2.** Attendance by socios and abonados, singles and married before (dark) and after (light) acts of violence and hooliganism.

time they become socios may not reflect accurately their status at the time they choose to go to a game. Still, during the past decades and even today, the probability that a single adult person will get married in Spain is much greater than the probability that a married person will get divorced. This empirical fact is something that generates a bias *against* the testable implication of the model. If singles are less likely to respond to acts of violence but more likely to be married (than a married person to get divorced), then mistakenly considering a married person as a single mistakenly means that we should tend to see similar reactions. Furthermore, for abonados the data should capture much more accurately their marital status, given that they have to get a new abonado pass every year.[1]

Consider now just abonados and the probability that they will renew their annual pass for the following season. Using their data we can evaluate:

*Hypothesis 3. Acts of violence, hooliganism, and vandalism during a season imply that the likelihood that an abonado will renew his or her pass for the following season is greater for singles than for married individuals.*

This hypothesis appears to be strongly confirmed by the aggregate data, with a more than 40% difference in the renewal rate between these two pools (see figure 9.3).

Finally, consider education. Individuals with greater cognitive (and noncognitive) skills assess objective risk more accurately and therefore face lower cost of overcoming fear. If, as suggested by the economics

1 It is true of course that they can still get married during the season. This fact means that data are still subject to the same bias, although likely much less strongly.

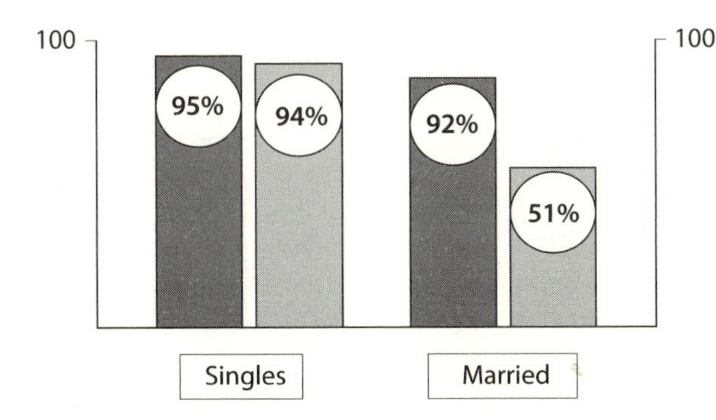


**Figure 9.3.** Renewal rate of season ticket for abonados, single and married, before (dark) and after (light) seasons that experienced acts of violence and hooliganism.

and psychology literature, we proxy these skills by education level, we can study the following hypothesis:

*Hypothesis 4. Socios and abonados with greater cognitive ability (proxied by education) should respond less to acts of violence, hooliganism, and vandalism than those with lower cognitive ability.*

Here we simply divide the education level on the ID card and on the application form into two types of professions (also indicated in these documents): those white collar professions that require a 5- or 6-year college degree (e.g., doctors, economists, lawyers, and engineers) and those that require just a high school degree or no degree (e.g., plumber, carpenter, and other blue collar jobs). The data are again consistent with the predicted differential response by education levels (see figure 9.4).

Finally, we take a look at the media. Media coverage of violence may trigger anxiety, distress, and fear. In the Becker–Rubinstein model, wider coverage lowers the expected marginal utility of $x$ in individuals who choose not to invest, whereas it has no effect on those who choose to invest. Moreover, if media coverage exaggerates *subjective* beliefs, it increases the economic incentives to invest in overcoming fear. Although media coverage of violence reduces consumption of $x$ for those who do not control fear, it might mitigate the effect on the consumption of $x$ for those who adjust their investment in controlling fear and hence bring their subjective beliefs more in line with objective dangers.

We study this aspect by focusing on games that ended after midnight and games that ended before midnight. During the 1950s to the 1980s, newspaper offices used to close around midnight, and journalists and photographers had to walk from the stadium to the office and rush to write their articles and print their pictures. This phenomenon means that

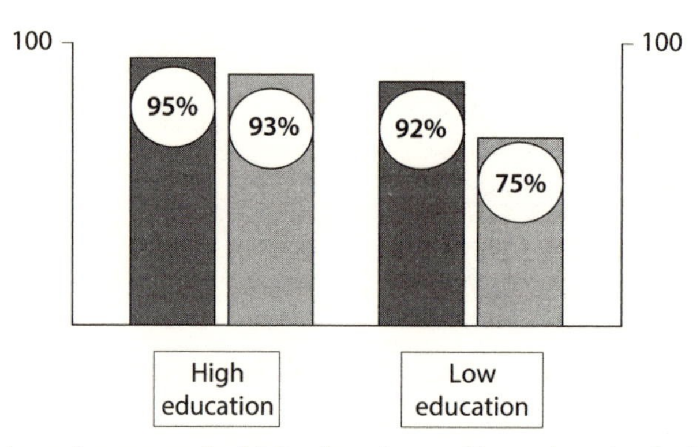


**Figure 9.4.** Attendance rate for high education and low education before (dark) and after (light) acts of violence and hooliganism.

games that started before 10:00 at night would likely allow the media to provide much more accurate, detailed, and descriptive evidence of the violence in the stadium, including pictures. Using this simple split of the games, it is possible to examine the following hypothesis:

*Hypothesis 5. The response to acts of violence, hooliganism, and vandalism should be lower after games that end after midnight than after games that end before midnight.*

Although there are not many games that started at 10:00 p.m. or later, or cup games that went into extra time and concluded close to midnight (less than 10% of the sample) and the data have to be taken with caution, the basic evidence shows a clear difference in attendance at the following home game that is again consistent with the hypothesis of the model (see figure 9.5).

*

Although a much richer data set would have been ideal to conduct precise econometric analysis, the basic evidence from these violent events supports the implications of the Becker–Rubinstein model in a number of unusual dimensions. It is not the risk of physical harm that moves people; it is the emotional disquiet. People respond to fear, not risk.

Interestingly, given the literature on favoritism under social pressure discussed in chapter 7 influencing the outcome of competitive sports, there is a surprising lack of research on the effects of crowd violence as a potent form of social pressure on the sporting and business side of sports. A recent article by Jewell et al. (2013) fills this gap. It studies the top four divisions of professional English soccer during an early period (1984 to 1994), in which hooliganism was a fundamental social problem,

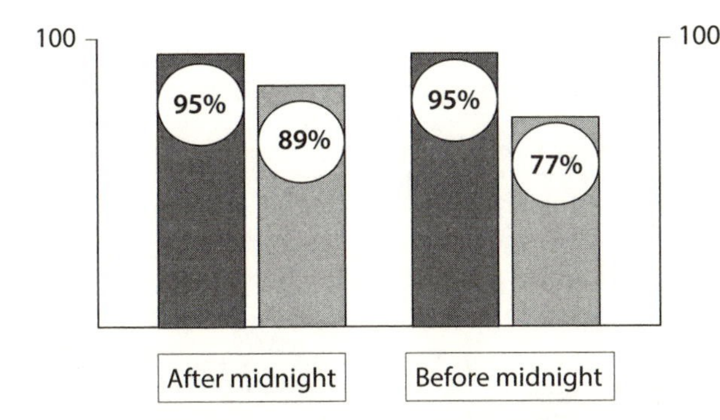


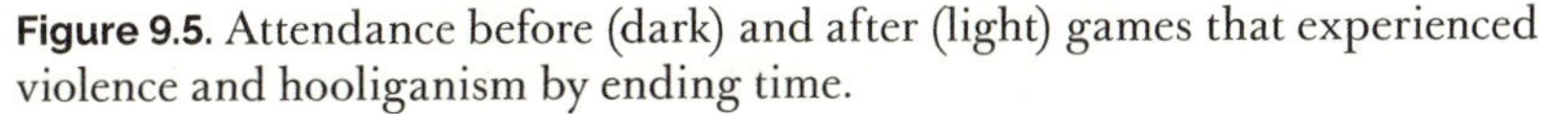


**Figure 9.5.** Attendance before (dark) and after (light) games that experienced violence and hooliganism by ending time.

and a more recent period (1999 to 2009), in which hooliganism has been less prevalent. Consistent with the results in chapter 7 and in this chapter, in the early period hooliganism has a *positive* effect on league performance and an *adverse* effect on soccer club revenues, whereas both of these effects disappear in the more recent period.

Besides adding to the recent literature on rational fear, the results in this chapter may also have implications for what is known as *crowd psychology*. This is a branch of social psychology that develops different theories for explaining the ways in which the psychology of the crowd differs from and interacts with that of the individuals within it. Dating at least as far back as Sigmund Freud, this field relates to the behaviors and thought processes of both the individual crowd members and the crowd as an entity. The recent Occupy movement, the London riots, and many other crowd events show the ongoing need for study of crowd phenomena. The model described and tested in this chapter appears to have a great deal of promise for what clearly is an important field of ongoing research.

# 10

# FROM ARGENTINA WITHOUT EMOTIONS

© TAWNG/STOCKFRESH

Psychologists and others in recent years have placed great emphasis on limits on individual rationality, but people train themselves to reduce and sometimes more than fully overcome their emotions.
—Gary Becker 1996

People are both emotional and rational. They have passions and they have interests (Hirschman 1977). It may not come as a big surprise, but in scientific research, it is often not trivial to delineate between rationality and emotions. Although the rationality principle provides excellent guidance to understand human acts, it is sometimes difficult to know when we can expect one type of behavior or the other. The analysis in the previous chapter shows that people exhibit both innate emotional responses to certain events and some capacity to control their emotions and limit their overreactions; the analysis also showed that the responses depend, at least in part, on the costs and the benefits.

The analysis in this chapter moves from fear to a different emotion. It returns to the emotion documented in chapter 5: psychological pressure in a competitive environment. That chapter showed how the pressure associated with the state of the competition (the leading–lagging asymmetry caused by the randomly determined order) in a dynamic tournament created a significant difference in the performance of the competitors. This effect was found among professionals in a high-stakes competitive setting. Thus, professionalism, high stakes, and competition did not appear to eliminate their emotional responses. Nevertheless, it must be noted that the vast majority of professional players rarely encounter the situation that was studied (a penalty shoot-out). Hence, it is not known what would happen if they had to face that situation much more frequently. If these subjects have some capacity to control their emotions and if the incentives to do so become sufficiently large, then, through training, practice, and other forms of investment, they should become less sensitive to psychological pressure. Put differently, if incentives become sufficiently large, their performance should become much less subject to the effects of anxiety, distress, and other pressures, perhaps even entirely free from psychological biases.

As in the previous chapter, this is a testable implication. This chapter takes advantage of a unique natural experiment in Argentina that was run for only one season to study this hypothesis in the same setting (a penalty shoot-out) as in chapter 5.

In the season 1988–89, the Argentine league championship decided to experiment with an unusual point system: After each drawn (tied) match, there would be a penalty shoot-out to determine which team got a bonus point. That is, the modified point system just for that season was as follows:

3 points if a team wins the match;
2 points if a team draws (ties) the match and wins the penalty shoot-out;
1 point if a team draws the match and loses the penalty shoot-out; and
0 points if a team loses the match.

The league had 20 teams, which played each other twice (at home and away) during the season. Therefore, a given team played 38 games in the season, roughly one per week, and the season had a total of 380 games played from September 1988 to May 1989. Given the relative frequency of a draw in Argentina (and most soccer leagues) in the 1980s of around 35–38%, the expected number of draws in the season was around 135–145 (if, of course, the experiment did not change the teams' behavior substantially). That season featured 132 draws (34.73%) and 131 shoot-outs.[1]

Most professional players in most countries, including those studied in chapter 5, are involved in no more than one or two penalty shoot-outs in their *lifetimes*. Assume with a substantial degree of generosity that a player expects to be involved on average in one shoot-out, say, every five years. The natural experiment in Argentina meant that in that season, every team was expected to be involved in roughly one shoot-out every *three weeks* (35% of all the games). That is, players were going to be exposed to a situation much more frequently and in a much shorter period of time than ever in the history of soccer. No team, country, league, or tournament has ever before or after featured such a high chance to be in that specific situation (a penalty shoot-out) for such a long period of time (every week for 38 weeks). In relative terms, this is close to a 9,000% (or 90 times) increase in the expected frequency of that situation. You'd better overcome any anxieties if you can! Thus, this situation provides a unique opportunity to study if the increase in incentives was sufficiently large to induce subjects, through training, practice, and other forms of investment in mental capital, to overcome their psychological pressure.

The basic data set for the experiment is available from well-known sources.[2] What is not often available is the key information necessary to evaluate our hypothesis: the kicking order. This information is reported in table 10.1 for all the 132 drawn games in the season. In addition to the round, the date, and the names of the teams, an asterisk (*) indicates the team that won the coin toss and therefore began kicking in the shoot-out, and in brackets are the final scores in the shoot-outs.

Before testing the hypothesis that the winning proportions for the first and second kicking teams are 50–50, it is worth noting that this data set is unique in that includes what for a long time was the answer to a trivia question posed in chapter 5: What is the world's longest penalty shoot-out? On November 20, 1988, the match between Argentinos Juniors and Racing Club finished 2–2, and 44 penalties were taken before Argentinos emerged as the 20–19 victors. On January 23, 2005,

1 There was no shoot-out in the game between Racing Club and Boca Juniors on December 22, 1988, which was suspended at halftime. Later, it was ruled that Racing Club would get no points and Boca Juniors would be awarded two points.

2 The complete data set, except the kicking order in the shoot-outs, is available at http://www.rsssf.com/tablesa/arg89.html.

**Table 10.1.** Information for the 132 Drawn Games in the 1988–89 Season

| Round | Date | Teams | Score |
|---|---|---|---|
| 1 | Sept. 11, 1988 | Ferro Carril Oeste–Newell's Old Boys* | [3]0–0[1] |
| | | Talleres de Córdoba–Mandiyú* | [4]1–1[2] |
| 2 | Sept. 14 and 18, 1988 | Mandiyú*–Rosario Central | [3]1–1[1] |
| | | Deportivo Español–Independiente* | [11]2–2[12] |
| | | Newell's Old Boys*–Instituto de Córdoba | [6]1–1[5] |
| 3 | Sept. 25, 1988 | Instituto de Córdoba–Gimnasia y Esgrima* | [4]0–0[5] |
| | | Ferro Carril Oeste–Racing Club* | [2]1–1[4] |
| | | Racing de Córdoba*–Deportivo Armenio | [2]0–0[3] |
| | | Boca Juniors*–Vélez Sarsfield | [2]0–0[3] |
| | | Platense*–Deportivo Español | [3]1–1[4] |
| | | Independiente–Argentinos Juniors* | [5]1–1[6] |
| | | Estudiantes La Plata*–Mandiyú | [4]1–1[5] |
| | | Rosario Central–Talleres de Córdoba* | [2]1–1[1] |
| 4 | Oct. 1 and 2, 1988 | Talleres de Córdoba–Estudiantes La Plata* | [5]2–2[4] |
| | | Deportivo Armenio–Ferro Carril Oeste* | [1]0–0[2] |
| 5 | Oct. 16, 1988 | San Martín de Tucumán*–Gimnasia y Esgrima | [3]1–1[4] |
| | | Newell's Old Boys–Racing Club* | [3]1–1[4] |
| | | Ferro Carril Oeste*–River Plate | [4]1–1[1] |
| | | Estudiantes La Plata–Rosario Central* | [5]0–0[4] |
| 6 | Oct. 23, 1988 | Argentinos Juniors–San Lorenzo* | [1]1–1[3] |
| 7 | Oct. 26, 30, and 31, 1988 | Racing de Córdoba*–Argentinos Juniors | [5]2–2[3] |
| | | Newell's Old Boys–River Plate* | [4]0–0[2] |
| | | Gimnasia y Esgrima*–Deportivo Armenio | [3]0–0[4] |
| | | San Lorenzo*–Mandiyú | [3]1–1[1] |
| 8 | Nov. 3, 1988 | Independiente–San Martín de Tucumán* | [3]0–0[2] |
| | | Estudiantes La Plata*–Platense | [5]0–0[3] |
| | | Talleres de Córdoba–San Lorenzo* | [3]1–1[5] |
| | | River Plate*–Gimnasia y Esgrima | [5]1–1[6] |
| 9 | Nov. 6, 1988 | San Martín de Tucumán–Deportivo Armenio* | [5]0–0[6] |
| | | Newell's Old Boys–Deportivo Español* | [3]1–1[1] |
| | | Ferro Carril Oeste*–Mandiyú | [4]0–0[2] |
| | | Racing de Córdoba*–Talleres de Córdoba | [4]0–0[2] |
| | | Platense*–Independiente | [4]0–0[2] |
| 10 | Nov. 12 and 13, 1988 | Rosario Central–Racing de Córdoba* | [2]0–0[4] |
| | | Estudiantes La Plata*–San Lorenzo | [2]1–1[3] |
| | | Argentinos Juniors*–Newell's Old Boys | [4]0–0[3] |
| | | Deportivo Español*–Gimnasia y Esgrima | [4]0–0[2] |
| 11 | Nov. 16, 17, and 23, 1988 | Instituto de Córdoba*–Talleres de Córdoba | [3]1–1[4] |
| | | Deportivo Armenio–Vélez Sarsfield* | [2]1–1[3] |
| | | Gimnasia y Esgrima*–Argentinos Juniors | [4]1–1[3] |
| | | San Lorenzo–Independiente* | [6]2–2[5] |
| 12 | Nov. 20, 1988 | Mandiyú*–Gimnasia y Esgrima | [2]0–0[4] |
| | | Argentinos Juniors–Racing Club* | [20]2–2[19] |

*(continued)*

**Table 10.1.** (*Continued*)

| Round | Date | Teams | Score |
|---|---|---|---|
| 13 | Nov. 27, 1988 | San Martín de Tucumán-Vélez Sarsfield* | [1]1-1[3] |
| | | Gimnasia y Esgrima-Talleres de Córdoba* | [4]0-0[3] |
| | | San Lorenzo-Boca Juniors* | [5]1-1[6] |
| 14 | Nov. 30, 1988 | Platense-Ferro Carril Oeste* | [7]1-1[6] |
| | | Independiente-Instituto de Córdoba* | [4]1-1[3] |
| | | Mandiyú*-Deportivo Armenio | [2]0-0[4] |
| | | Deportivo Español-Vélez Sarsfield* | [5]2-2[6] |
| 15 | Dec. 5 and 6, 1988 | San Martín de Tucumán*-Deportivo Español | [2]0-0[4] |
| | | Vélez Sarsfield-Argentinos Juniors* | [4]1-1[3] |
| | | River Plate-Mandiyú* | [4]4-4[2] |
| | | Racing Club*-Rosario Central | [3]0-0[4] |
| | | Newell's Old Boys*-Independiente | [5]2-2[3] |
| | | Ferro Carril Oeste-Boca Juniors* | [5]0-0[6] |
| 16 | Dec. 14, 1988 | Boca Juniors*-Instituto de Córdoba | [3]1-1[2] |
| | | Platense-Newell's Old Boys* | [5]0-0[4] |
| | | Estudiantes La Plata-Racing Club* | [2]0-0[3] |
| | | Rosario Central*-Deportivo Armenio | [5]3-3[3] |
| | | Mandiyú*-Vélez Sarsfield | [5]1-1[3] |
| | | Argentinos Juniors-Deportivo Español* | [4]0-0[2] |
| 17 | Dec. 11, 1988 | San Martín de Tucumán- Argentinos Juniors* | [2]1-1[4] |
| | | Deportivo Español-Mandiyú* | [3]1-1[5] |
| | | Gimnasia y Esgrima*-Platense | [5]1-1[6] |
| | | Newell's Old Boys*-Boca Juniors | [4]1-1[5] |
| 18 | Dec. 18, 1988 | Ferro Carril Oeste-San Martín de Tucumán* | [4]0-0[5] |
| | | Platense*-Racing Club | [6]0-0[5] |
| | | Estudiantes La Plata*-River Plate | [6]1-1[5] |
| 19 | Dec. 22, 1988 | Deportivo Español-Rosario Central* | [4]1-1[3] |
| | | Vélez Sarsfield-Estudiantes La Plata* | [5]0-0[4] |
| | | Racing Club-Boca Juniors | 0-0 |
| 20 | Jan. 28, 1989 | Vélez Sarsfield*-Independiente | [3]0-0[4] |
| 21 | Feb. 4 and 5, 1989 | Racing de Córdoba*-Racing Club | [4]1-1[1] |
| | | Boca Juniors*-River Plate | [3]0-0[4] |
| | | San Lorenzo-Deportivo Armenio* | [3]0-0[2] |
| | | Instituto de Córdoba-Newell's Old Boys* | [3]1-1[4] |
| 22 | Feb. 11 and 12, 1989 | Talleres de Córdoba*-Rosario Central | [3]5-5[1] |
| | | Racing Club-Ferro Carril Oeste* | [4]1-1[2] |
| | | Mandiyú-Estudiantes La Plata* | [3]2-2[1] |
| 23 | Feb. 19, 1989 | San Martín de Tucumán*-Rosario Central | [1]0-0[3] |
| | | Estudiantes La Plata*-Talleres de Córdoba | [3]0-0[2] |
| | | Ferro Carril Oeste*-Deportivo Armenio | [5]1-1[4] |
| 24 | Feb. 26, 1989 | Vélez Sarsfield*-Racing de Córdoba | [3]1-1[2] |
| | | Talleres de Córdoba*-Independiente | [2]1-1[4] |
| 25 | March 5, 1989 | Racing de Córdoba-Deportivo Español* | [4]0-0[3] |
| | | Ferro Carril Oeste*-Vélez Sarsfield | [4]1-1[3] |
| | | Instituto de Córdoba*-River Plate | [4]0-0[5] |
| | | Gimnasia y Esgrima*-Racing Club | [4]1-1[2] |

**Table 10.1.** (*Continued*)

| Round | Date | Teams | Score |
|---|---|---|---|
| 26 | March 11 and 12, 1989 | Racing Club–San Martín de Tucumán* | [4]0–0[3] |
| | | Deportivo Armenio*–Gimnasia y Esgrima | [2]0–0[3] |
| | | Argentinos Juniors*–Racing de Córdoba | [5]1–1[3] |
| | | Estudiantes La Plata–Independiente* | [2]1–1[4] |
| 27 | March 18 and 19, 1989 | Racing de Córdoba*–Mandiyú | [1]0–0[3] |
| | | Ferro Carril Oeste–Argentinos Juniors* | [4]0–0[3] |
| | | Newell's Old Boys*–Vélez Sarsfield | [3]1–1[4] |
| | | Gimnasia y Esgrima–River Plate* | [5]0–0[4] |
| | | Racing Club–Deportivo Armenio* | [3]2–2[5] |
| 28 | March 25 and 26, 1989 | Rosario Central–San Lorenzo* | [3]0–0[1] |
| | | Deportivo Español*–Newell's Old Boys | [4]0–0[5] |
| | | Mandiyú–Ferro Carril Oeste* | [3]0–0[1] |
| 29 | April 2, 1989 | Newell's Old Boys*–Argentinos Juniors | [3]0–0[2] |
| | | Racing Club*–Vélez Sarsfield | [3]1–1[4] |
| | | Deportivo Armenio*–River Plate | [4]1–1[2] |
| 30 | April 9 and 10, 1989 | Vélez Sarsfield–Deportivo Armenio* | [3]1–1[2] |
| | | Platense*–Boca Juniors | [5]1–1[6] |
| 31 | April 16 and 17, 1989 | Gimnasia y Esgrima–Mandiyú* | [3]1–1[1] |
| | | Instituto de Córdoba–Rosario Central* | [5]2–2[4] |
| | | San Lorenzo–Platense* | [3]0–0[1] |
| 32 | April 23, 1989 | Deportivo Español–River Plate* | [3]1–1[4] |
| | | Mandiyú*–Racing Club | [5]1–1[4] |
| | | Platense*–Racing de Córdoba | [2]1–1[4] |
| 33 | April 30, 1989 | Gimnasia y Esgrima*–Rosario Central | [3]0–0[4] |
| | | Deportivo Armenio*–Mandiyú | [8]0–0[7] |
| | | River Plate–Argentinos Juniors* | [4]2–2[2] |
| | | Vélez Sarsfield–Deportivo Español* | [3]2–2[5] |
| 34 | May 6 and 7, 1989 | Rosario Central*–Racing Club | [3]2–2[4] |
| | | Argentinos Juniors*–Vélez Sarsfield | [2]1–1[4] |
| | | Mandiyú–River Plate* | [4]0–0[2] |
| | | Estudiantes La Plata–Gimnasia y Esgrima* | [0]1–1[2] |
| 35 | May 16, 1989 | San Martín de Tucumán–Racing de Córdoba* | [5]0–0[4] |
| | | Ferro Carril Oeste–San Lorenzo* | [7]1–1[8] |
| | | Vélez Sarsfield*–Mandiyú | [3]0–0[2] |
| 36 | May 21, 1989 | Talleres de Córdoba*–Vélez Sarsfield | [4]0–0[5] |
| | | Rosario Central*–River Plate | [2]1–1[4] |
| | | Independiente*–Racing Club | [4]0–0[1] |
| 37 | May 25, 1989 | Gimnasia y Esgrima–Boca Juniors* | [2]1–1[4] |
| | | Argentinos Juniors*–Mandiyú | [12]0–0[13] |
| 38 | May 28, 1989 | Talleres de Córdoba–Argentinos Juniors* | [10]0–0[9] |
| | | Rosario Central–Deportivo Español* | [1]1–1[4] |
| | | Platense–Deportivo Armenio* | [3]1–1[2] |
| | | Boca Juniors*–Racing Club | [4]1–1[3] |

[1] Game suspended at halftime with no shoot-out.

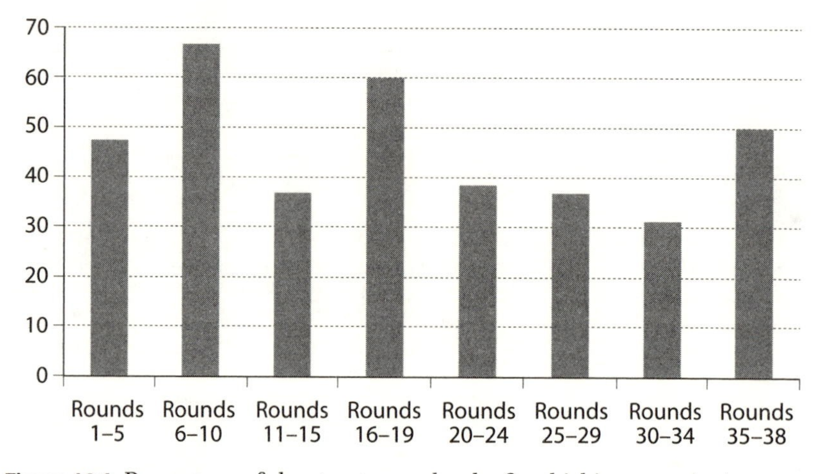


**Figure 10.1.** Percentage of shoot-outs won by the first kicking team in Argentina, 1988–89 season.

however, a new record was set in the Namibian Cup, known as the Tafel Lager NFA Cup. KK Palace played Civics in the first round. The final score was a 2–2 draw. In the ensuing penalty shoot-out, 48 penalties were taken before KK Palace emerged as the 17–16 victors.

The data show that 60 out of 131 shoot-outs (45.8%) were won by the team kicking first and 71 by the second-kicking team (54.2%) (see figure 10.1). Given the sample size, the hypothesis that the winning proportions are 50–50 cannot be rejected at conventional levels ($p$-value = 0.382). Perhaps in the first couple of months of the competition (10 rounds) there appears to be a small effect favoring the first team (it won 21 out 37 shoot-outs, or 56.7%), but it is statistically insignificant. Actually, if anything, there seems to be an effect going in the opposite direction (favoring the second-kicking team) after those rounds and especially later in the season, but again it is statistically insignificant.

Thus, as expected from the results in the previous chapter and the large change in incentives, there is no evidence in the aggregate data that players were subject to psychological pressure in the shoot-outs during the season. Interestingly, there is also little evidence that specific teams performed significantly better or worse than 50–50, at most a couple of teams at standard significance levels (see table 10.2).[3]

In conclusion, the evidence is consistent with the hypothesis that professionals belonging to the *same* pool of subjects that exhibited

3 Care must be exercised in the statistical testing because the winning proportions for the teams are not independent. When one team wins a shoot-out, another loses.

**Table 10.2.** Shoot-Outs Won and Lost by Team

| | | Shoot-Outs | |
|---|---|---|---|
| Team | Matches Drawn | Won | Lost |
| Vélez Sarsfield | 17 | 12 | 5 |
| Gimnasia y Esgrima | 16 | 11 | 5 |
| Deportivo Mandiyú | 19 | 11 | 8 |
| San Lorenzo | 10 | 8 | 2 |
| Boca Juniors* | 8 | 6 | 2 |
| Independiente | 11 | 7 | 4 |
| Newell's Old Boys | 13 | 7 | 6 |
| Ferro Carril Oeste | 14 | 7 | 7 |
| Deportivo Armenio | 15 | 7 | 8 |
| Rosario Central | 16 | 7 | 9 |
| Platense | 11 | 6 | 5 |
| Racing de Córdoba | 11 | 6 | 5 |
| River Plate | 13 | 6 | 7 |
| Deportivo Español | 14 | 6 | 8 |
| Racing Club* | 15 | 6 | 9 |
| Argentinos Juniors | 16 | 6 | 10 |
| Talleres de Córdoba | 12 | 5 | 7 |
| Estudiantes La Plata | 12 | 4 | 8 |
| San Martín de Tucumán | 10 | 2 | 8 |
| Instituto de Córdoba | 9 | 1 | 8 |

*This table does not include the game played on December 22 (Round 19) between Racing Club and Boca Juniors.

psychological pressure in chapter 5 were able to overcome their emotions through training, practice, and other investments in mental capital.

As a reasoned deduction or inference, however, this conclusion may appear to be somewhat speculative at this point. It seems fine to consider that the players belong to the *same* pool of subjects as in previous chapters. In fact, there is statistical evidence that supports this point. Econometric regressions with fixed effects for nationality that try to explain the scoring rate in penalties during the course of a game (chapter 1) and in penalties kicked in a shoot-out (chapter 5) are statistically insignificant at standard confidence levels. In other words, Argentineans are not different from players from other nationalities at penalty kicks. What seems a stretch is to argue that the mechanism that helped these players overcome the pressure in a penalty shoot-out definitely was through training, practice, and other forms of investment in mental capital rather than, say, through drinking more *mate* (sometimes called the Argentinians' national drink). More evidence is needed.

**Table 10.3.** Survey of Training Frequencies for Penalty Shoot-Outs

| Country | *N* | During League Competitions (times per week) | During Elimination Tournaments (times per week) |
|---|---|---|---|
| Argentina (1988–89 season) | 25 | 4.272 | 4.272 |
| Argentina (other seasons) | 27 | 0 | 0.033 |
| England (any season) | 38 | 0 | 0.011 |
| Germany (any season) | 40 | 0 | 0.117 |
| Italy (any season) | 33 | 0 | 0.070 |
| Spain (any season) | 42 | 0 | 0.052 |

*Note*: *N* is the number of coaches and players in the survey who participated in the competitions representing a team from the corresponding country. Elimination tournaments include national cup and international cups for clubs. The sample is too small to include the World Cup or European Cup.

The following survey addresses this point. It asks professional coaches and players who participated in the Argentinean League before, during, and after that 1988–89 season about their training practices in penalty shoot-outs. It also asks coaches and players from other leagues and countries in various years. Unsurprisingly, the results show a huge difference in training frequencies in Argentina in the 1988–89 season with respect to all the other countries and years, including Argentina in other seasons (see table 10.3). It probably was not drinking *mate* after all, but rather the intense weekly training (about 70 times greater) that helped them reduce and more than fully overcome their emotions.

"I've only taken one penalty before, for Crystal Palace at Ipswich. It was 2-2 in the eighty-ninth minute. I hit the post and we went down to second division that year. But I think I'd be far more comfortable now than I was then." This is Gareth Southgate, English international, during the early stages of Euro '96. There is no evidence that he had practiced penalty kicks more often than before. A few days later, his miss saw England suffer the misery of a penalty shoot-out defeat by Germany in the semifinals.

*

The systematic analysis of individual responses to changes in the environment is important for understanding the determinants of emotions and the extent and formation of rationality. The fact that observed behavior in the 1988–89 season in Argentina, where the frequency and incentives are substantially greater than in the standard setting, is different from that in chapter 5, strongly suggests that people can train themselves to reduce and sometimes fully overcome their emotions.

It may be just a coincidence (the sample is small to tell), but on June 30, one year and a month after this experiment concluded, Argentina beat Yugoslavia 3–2 in a penalty shoot-out to qualify for the semifinal in the 1990 FIFA World Cup in Italy. Argentina was the first team to kick. Three days later, Argentina beat the hosts Italy 4–3 in the semifinal, again in a shoot-out, to qualify for the World Cup final. This time, Argentina kicked second. Throughout the World Cup, and even in the final, Argentina was often reported to seem to "play for penalties," thanks in part to a terrific goalkeeper saving penalties, Sergio Goycochea. Obviously, it can only ever be anecdotal, but the success rate and the confidence in penalty shoot-outs that Argentina exhibited in that World Cup seems to be a perhaps unintended benefit of the experiment one year earlier.

11

# DISCRIMINATION: FROM THE MAKANA FOOTBALL ASSOCIATION TO EUROPE

© TAWNG/STOCKFRESH

Sport has the power to change the world. It has the power to inspire, the power to unite people that little else has. It is more powerful than governments in breaking down racial barriers.

—Nelson Mandela, giving a lifetime achievement award to Pelé

Mandela knew better than anyone that sports represent an excellent setting to document racial differences. For decades, he witnessed black inmates playing soccer in Robben Island at the Makana Football

Association while whites played rugby in the rest of the country. If his intuition was right, sports would also be extremely valuable to break down racial differences. With the benefit of hindsight, it is no surprise that he picked sports to try to do just that. And among sports, he picked one of the more far-fetched causes imaginable: the white people's sport (rugby) and the national rugby team (the Springboks) who would host the sport's World Cup in 1995 and were the embodiment of white supremacist rule during apartheid.

In his best-selling book *Playing the Enemy* (2008), John Carlin describes how the Springboks capped Mandela's miraculous 10-year-long effort to bring South Africans together. One of the opening shots of Clint Eastwood's movie *Invictus*, based on this book, focuses on two groups of children playing across the street from each other. On one side, big brawny white boys toss the rugby ball back and forth in full Afrikaner school uniform on a perfectly manicured field. Across the road, skinny, tatttered black children kick around a soccer ball in a cloud of dust. Mandela, recently released from jail, drives by to the utter delight of the black children, who scream his name with faces pressed up against the fence. The Afrikaners stand silently by.

*

In the past two decades, substantial progress has been made in the development of theoretical frameworks to study different aspects of racial differentials in wages, employment, and other labor market outcomes. Lang and Lehmann (2012) offer an excellent review of the literature. The different theoretical models of discrimination are typically divided into three categories: models based on a "taste" for discrimination (where some individuals gain utility from discriminating behavior, e.g., racial discrimination), models of statistical discrimination (in a world of imperfect information, market participants might use race as a proxy for an unobservable characteristic they are interested in), and models of social interactions (a game theoretic explanation of discrimination in terms of unintended aggregate consequences of individual behavior, as in Schelling (1971)).

This chapter is concerned with taste-based models of discrimination, where the pioneer framework of analysis is the model first formalized in Gary S. Becker's classic *The Economics of Discrimination* (1957).

The Becker model of discrimination differs from almost all other major models of discrimination in that it departs from the standard assumption that firms maximize profits or very nearly so. As long as discrimination persists in equilibrium, prejudiced firms earn lower profits. Similarly, if workers engage in prejudice (e.g., by refusing to work with certain groups of workers) or if consumers do (e.g., boycotting products supplied by those groups), then they forego earnings or pay higher prices because of their prejudice.

A common empirical approach to test these models is to test directly whether the wage equals the value of marginal product, as predicted by profit maximization. We can test, for instance, whether subjects with similar statistical measures of performance receive equal wages. Such studies, however, are always subject to the criticism that the statistical measures we observe may not capture productivity exactly. It is possible to think of estimating a production function and ask whether wage differentials are proportional to marginal productivity differences. This approach, however, requires strong and implausible assumptions to identify the marginal product (e.g., the distribution of worker types is assumed to be exogenous rather than endogenous; the distribution of jobs is assumed to be independent, rather than dependent, of race, and others).

Szymanski (2000) suggests a novel test using soccer. Assume that soccer fans care only about winning and that team owners care only about pleasing the fans and thereby increasing profits. Then, for a given salary bill, the teams' win–loss records should be *independent* of the racial composition of the team. If, among teams with the same total salary bill, those with more black players have better records, then a team could improve its performance by hiring more blacks. Either the team owners must not be maximizing profits, or consumers must care about the racial composition of the team. This is the model we study in this chapter.

Taste for discrimination models can be distinguished with regard to which type of agent in the market is assumed to have a liking for discrimination: owners, employees, or customers (see, for instance, Kahn (2000)). If owners discriminate, the consequences in the long run are relatively benign: As long as there exist some nondiscriminating owners, for reasons we will see later, nondiscriminating owners eventually drive the discriminators out of the market so that discrimination does not persist forever. If employees discriminate, the result is firms that are racially monolithic, but under a certain set of plausible assumptions, there is no inefficiency in the market. Finally, if customers have a taste for discrimination, it is in some sense efficient for the market to supply services that cater to these tastes and, therefore, discrimination is not eliminated, even in the long run. Of course, this is a theory and, as such, it is unclear whether it is likely to be observed in reality. We need to do empirical work.

But this empirical work is not easy. Generally, the conventional approach to testing for racial discrimination is to specify an earnings function, that is, a relationship that describes wage income as a function of personal characteristics that influence productivity like education, work experience, or proxies for intelligence. Using this function, any systematic differences in earnings between racial groups, after controlling for all factors relevant for earnings, are taken as evidence of racial discrimination. As is well known in econometrics, this estimation procedure suffers from a fatal flaw: If any of the relevant factors is not observable and differs on average among the different ethnicities, then the model suffers from the so-called *omitted variable bias*. Suppose, for instance, that the cultural heritage of an ethnic group strongly emphasizes hard work and autonomous thinking (qualities that are also highly valued by employers). Then, traditional econometric models would attribute differences in earnings that are caused by the difference in work ethic to racial discrimination, rather than to the differential work ethic or cultural heritage of the different ethnic groups. Given the pervasiveness of unobserved variables, this bias is clearly a crippling problem for the empirical investigation of racial discrimination.

In 2000, Stefan Szymanski found a clever way out of this problem: Instead of worrying endlessly (and probably fruitlessly) about the omitted variable bias, he argued for a market-based test of racial discrimination. Suppose that every worker has two attributes: a level of talent that is identifiable (and can vary among workers) and an ethnic group. Suppose that there is an efficient labor market, and, further, that firms' profits are positively related to workers' abilities. Assume also that some employers have a "taste for discrimination," that is, in addition to any monetary profits, they also earn a "psychic profit" from hiring workers of their preferred ethnic group. Then, if the share of discriminating employers is high enough, this setup allows us to test directly for discrimination in the market: Workers from the discriminated-against ethnic group will be less in demand, ceteris paribus, and therefore will earn a lower equilibrium wage. Nondiscriminating firms could then achieve the same revenue at a lower cost by hiring predominantly from the excluded group, while discriminating firms would be compensated for their lower monetary profits through utility derived from their discriminating behavior.

Therefore, it may be possible to perform a "market test" for discrimination if we can find a setting where performance is easily measurable, workers' ability is a very important source of firm performance, and the labor market is efficient. As in the other chapters, this setting exists in soccer.

In England, soccer clubs are operated as business firms, owned privately or publicly by shareholders and filing annual financial statements accessible to everyone for inspection. A club's main sources of income are ticket sales and TV revenues (although recently, sponsorship contracts are becoming increasingly important), and costs primarily stem from the player payroll. The market for soccer players is international and highly competitive, with players often switching between countries and changing clubs for significant transfer fees. According to the management consultancy Deloitte (2013), in 2012–13, for instance, the clubs in the highest English division alone spent ≈600 million euros on transfer fees for new players, with net transfer costs of ≈200 million euros.

The evidence in this chapter comes from English league soccer, and so it is important to present a few relevant features of the English soccer system.

Firstly, as in most countries, league competition in English soccer is hierarchical, with four divisions, each containing about 20 teams. The highest division is the English Premier League (EPL). At the end of each season, the highest ranked teams from each lower division are promoted to the next higher division, and the teams with the least points are relegated into the next lower division. As opposed to US team sports, there are basically no play-offs, and hence competition is solely about the position in the league. Teams play about 38 league matches per season, which is enough for us to be fairly confident that performance is caused by underlying talent and not simply by chance events.

Secondly, redistribution of income between teams is limited, and especially when compared to professional sports in the United States. Moreover, other methods to maintain a competitive balance between teams that are popular in the United States (e.g., draft picks or salary caps) are also not used. Therefore, the financial success of a club is tied much more closely to its own performance than is the case in the United States.

Thirdly and very importantly, the market for professional football players in England is free from regulatory constraints, effectively since 1978. No collective bargaining over salaries exists, and neither do salary caps. Moreover, clubs trade professional footballers between each other frequently. For example, according to the website transfermarkt.de, in 2012–13 the total number of squad players in the English Premier League was 533, and in the same season the total number of players transferred in was 344 and transferred out was 304. However, a large fraction of these players were being loaned out to clubs in lower divisions, a practice that was less common in the period covered by Szymanski's data. But there has always been substantial mobility.

Finally, because the density of clubs in most parts of Great Britain is far higher than it is in the United States, the level of competition between professional sports teams for a loyal fan base and lucrative sponsorship contracts is also much more intense. To give just one example of this density, within 100 miles of Manchester United, arguably one of the most popular soccer clubs in the world (and definitely one of the most successful ones in English soccer in recent decades), there are about 50 other professional soccer clubs vying for the attention of spectators and sponsors.

An implication of the last two characteristics of the EPL is that if both player and fan markets are competitive, professional soccer players should earn wages that reflect their marginal product. And thus a club's spending on player wages should be a reasonably good indicator of competitive success on the pitch. Therefore, by backward induction, if the performance of English clubs turns out to be highly correlated with their wage bills, this will represent evidence consistent with the idea that the market for players is competitive, which is an assumption that is needed to derive our testable hypothesis.

The evidence, in fact, shows a strong relationship. Plotting a club's average wage bill with its corresponding league rank is visually quite persuasive, and formally, if we ran a simple regression of league position on clubs' wage expenditure, we obtain an $R^2$ about 0.90 (see figure 11.1).

Naturally, to implement the test of market-based discrimination, we require there to be not only a competitive labor market but also a significant share of black players in the English soccer leagues.

Although the Roman invasion of the British Isles and England's later position as the world's leading trading nation must have provided for an early presence of black people, the black population in England did

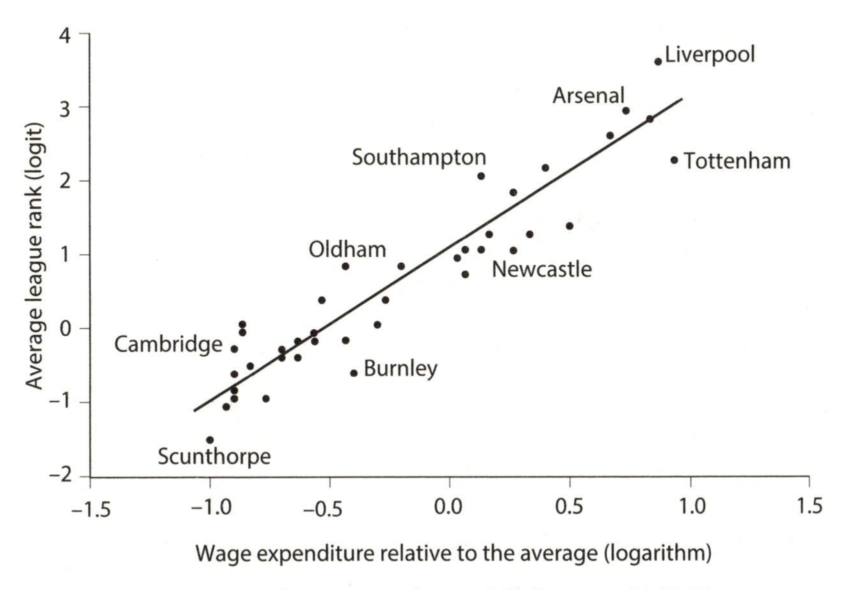


**Figure 11.1.** Performance and wage bill (Szymanski 2000).

not become a significant share of the total population until the 1950s. Around that time, immigrants (mostly from South Asia and the Caribbean, but also from China and Africa) were invited into the country by successive governments in an attempt to solve the chronic labor shortages that plagued the United Kingdom in those days. As a result of this migration, around 900,000 UK residents describe themselves as black (around 1.6% of the total population), according to the 1991 official UK census. However, what is relevant for our test is not the proportion of blacks in the total UK population, but in the subset of professional soccer players.

It is important to stress that even with this substantial migration, Britain is far less ethnically diverse than the United States. During the period, most people would have identified two different ethnic immigrant groups, those from the Indian subcontinent and those from the Caribbean. These two groups imported very different cultures. In sporting terms, the subcontinental migrants mostly played and were interested in cricket. Those migrants, and their descendants, who started to affect soccer from the early 1970s, were from the Caribbean and were most commonly referred to as "black."

For this research, a list of black players was compiled from inspection of team photographs, where players are classified as "black" if they look "black." This method might sound arbitrary, but it is in fact a good way to model the potential for discrimination. To see why, consider the legendary Manchester United player Ryan Giggs. He started playing for the club in the 1990 and was still playing regularly in 2012–13, when this book was being written. It was only once he was well established in his career that he publicly discussed the fact that his father was black and that he had been the subject of racial taunts in school because of his father's skin color. This phenomenon came as a surprise to his fans, since he himself looks Caucasian. It was unlikely that he faced discrimination as a professional player, precisely because discriminators prejudge an individual based on appearances.

So Szymanski constructed a data set of black players in a sample of 39 (out of 92) professional clubs from player records during the period 1978–93, during which period the share of black players was already significant. He found that whereas in 1978 there were only 4 black players appearing a total of 77 times (matches played) for the sample clubs, by 1993 there were 98 players appearing 2,033 times. Because the average squad size was around 25 players, by 1993 about 10% of all players were black.

The variation in the share of black players between clubs during 1978–93 is also considerable. The number of appearances or matches played by black soccer players ranges from 2 for the club with the least black player appearances in the whole sample to more than 1,000 for four clubs. Two football clubs among those sampled had fewer than 10 appearances, and four had fewer than 100.[1] (Trivia tidbit: The first black player to play on the national team of England was Viv Anderson on November 29, 1978, in Wembley against Czechoslovakia in front of 92,000 spectators. England won 1–0. Before him, Laurie Cunningham had already represented England but on the under-21 team.)

Stripped to the bone, the idea is to analyze how the proportion of black players employed affects the success of English soccer clubs after controlling for wage expenditure using a data set of 39 clubs over the 16-year period 1978–93.

The formal model that guides this empirical work is the following. Suppose that a team owner $i$ maximizes a utility function that is some weighted average of financial profits, $\pi_i$, and the share of white players in the team $s_i$ (this setup models his or her taste for racial discrimination), which is defined as

$$s_i = t_{iw}/(t_{iw} + t_{ib})$$

where $t_{iw}$ and $t_{ib}$ are white ($w$) and black ($b$) talent on a team.[2] The objective function for team $i$ is then given by

$$\Omega_i = \alpha s_i + (1 - \alpha)\pi_i$$

Now, financial success, $\pi_i$, is defined as revenues (which in turn depends on a team's success on the pitch, $w_i$) minus costs, which are a function of the cost of the total quantity of playing talent, $T_i \equiv t_{iw} + t_{ib}$:

$$\pi_i = R_i(w_i(T_i)) - c(s_i)T_i$$

Competitive success on the pitch, $w_i$, may be measured in terms of the percentage of wins, which obviously would disadvantage teams in higher divisions (where competition is harder and at a higher level), or in terms of championship success, which would ignore the fact that among the teams that do not win, not all get equally close to winning. Therefore, given the structure of English professional soccer, it is most

1 A matching stratified sample of nonblack players by birth dates showed that black players enjoy significantly greater playing longevity and accumulate more league appearances than nonblack players. Black players during this period were also more likely to play in attack and less likely to play in defense than their nonblack peers and to score more goals, probably because of their playing positions. Finally, black players are more likely to represent their own countries than their white counterparts, a feature that might offer evidence for greater playing ability.

2 To take into account the fact that credit for a victory is usually attributed to players in proportion to their talent, it is assumed that discriminating team owners prefer white to black talent, that is, that they prefer having a white superstar over a black superstar and having a white substitute player rather than a black one. Therefore, the taste for discrimination in $s_i$ is modeled in terms of talent.

straightforward to simply use league positions to measure competitive success. Because of promotion and relegation, the teams in Szymanski's data set move among the four professional English soccer divisions. Thus, each team has a unique rank at the end of each season. The team at top of the highest division has the position 1; when there are 20 teams in the top division, then the bottom team has a rank of 20, and the top team from the next highest division has a rank of 21, and so on.

The model further assumes sporting success to be a function of the total quantity of sporting talent hired by the club, $T_i \equiv t_{iw} + t_{ib}$. In other words, although the definition of talent may differ depending on the position of a player, talent can be somehow aggregated and treated as a homogeneous input. This method also implies that talent does not differ between ethnicities (that is, talent is defined in the same way for blacks and whites), which in turn does not require us to make any assumptions on the actual distribution of talent within a given population of players (e.g., black players).

Because discriminatory owners require not only that their players have a desired level of talent but also that they possess an additional characteristic (that is, a given skin color), they generally face higher costs for hiring. In other words, a discriminatory behavior increases costs and lowers profits. Specifically, the model assumes that for a given intensity of discrimination, $\alpha$, talent is supplied at a constant marginal cost, and the exact marginal cost of talent to a club depends on its owners' tastes for discrimination:

$$\text{If } s_i > s^*, \text{ then } c_i(s_i) = c_{io}[1 + (s_i - s^*)^2]$$
$$\text{If } s_i \leq s^*, \text{ then } c_i(s_i) = c_{io}$$

Here, $s^*$ is defined as the share of white players employed in a team beyond which a club starts to pay a premium for white talent. In other words, as long as not too many team owners are discriminating, the price for black and white players is approximately equal. However, owners who have a taste for discrimination get utility from hiring white players above and beyond what they are worth in terms of talent. That is, they are willing to pay a premium for them, and the size of this premium depends on the strength of their taste for discrimination relative to the importance they attach to financial gains.

At the same time, discriminatory owners lower the cost of black talent to nondiscriminators by giving them, in effect, a certain degree of monopsony power. Suppose that teams are interested in hiring a player of a given talent, and a white player and a black player are available. The black player will only be hired by a discriminating team owner if his wage is so much lower that it offsets the additional utility the discriminating owner gets from hiring the white player. Nondiscriminatory teams can exploit this fact by offering the black player a wage just above their equilibrium wage in a scenario with only discriminating owners.

If the number of black soccer players in the market is small, owners may get away with a certain degree of racial discrimination without paying a premium. Intuitively, discriminating owners in this case face virtually the same talent pool as their nondiscriminating rivals, and therefore the premium for white talent is negligible. However, as we have seen above, the share of black players in the sample was already significant (especially in later years), and therefore it is likely that owners with a taste for discrimination will have to pay to indulge in it.

Thus, the argument is that the exact value of $s^*$ arises endogenously from the interplay of demand and supply, and that in the case of discrimination, $s^*$ is at least as great as the share of white players in the labor market.

From the above equations, we can find the optimality conditions for $t_{iw}$ and $t_{ib}$ in firm (team owner) $i$:

$$\alpha(\partial s_i/\partial t_{iw}) + (1-\alpha)[(\partial R_i/\partial w_{ib})(\partial w_i/\partial t_{ib}) - c_i'(\partial s_i/\partial t_{iw})T_i - c_i(\partial T_i/\partial t_{iw})] = 0$$
$$\alpha(\partial s_i/\partial t_{ib}) + (1-\alpha)[(\partial R_i/\partial w_{ib})(\partial w_i/\partial t_{ib}) - c_i'(\partial s_i/\partial t_{ib})T_i - c_i(\partial T_i/\partial t_{ib})] = 0$$

Subtracting one condition from the other, and noting that $(\partial s_i/\partial t_{iw}) - (\partial s_i/\partial t_{ib}) = 1/T$, the optimal share of white talent for a discriminating owner (that is, $\alpha > 0$) is given by

$$t_{iw} = s^*T + \alpha/2(1-\alpha)c_{io}$$

In other words, if an owner discriminates against black players, the share of white players on the team is greater than the share of white talent in the total population (that is, $s^*T$) and, as a result, he or she pays a premium for the white players' talent.

In the case of a nondiscriminating owner (that is, $\alpha = 0$), the optimal share of white playing talent in the team is indeterminate but is in any case no greater than $s^*$ (in other words, nondiscriminating owners never pay a premium for white talent).

Intuitively, in this model an owner's preference for discrimination can be simply seen as a tax on the success of his or her team. For a given level of talent, the club has to pay more than a nondiscriminating club.

This notion yields a testable hypothesis: The expected performance (in terms of team position in the English soccer leagues), given any level of wage expenditure, will be worse for teams of discriminating than nondiscriminating owners.

Therefore, if a taste for discrimination is behind the observed shares of black professional soccer players in some clubs, a regression of league position on the race variable, controlling for the wage bill, should yield a statistically significant coefficient. On the other hand, if factors completely uncorrelated with racial discrimination are responsible for the distribution of black soccer players, the coefficient of the race measure should not be statistically different from zero.

Note that all we have said so far hinges on the assumption of a competitive market for players. If the labor market for soccer players is competitive, this fact allows us to uncover systematic discrimination against subgroups of players as long as these groups can be distinguished by a feature that is observed by the researcher.

Formally, the regression model is the following. The league position, $p_{it}$, is assumed to depend on five explanatory variables.[3] First, the difference (in logs) between a club's wage bill and the annual average, $w_{it} - w_t$; second, as a control for "bad luck," the player turnover of a team relative to the annual average (a high turnover usually signals a lot of injuries and is unsettling for players in and of itself), $\text{play}_{it} - \text{play}_t$; third, divisional dummies, $D_{jit}$, which reflect the fact that teams can only move between divisions at the end of each season; fourth, club-specific fixed effects, $\alpha_i$, which can potentially affect performance (for instance, the passion of the supporters or local weather conditions) but can be eliminated via first-differencing; and, finally, a variable measuring the share of black players' appearances for a team in a given season relative to the annual average, $\text{black}_{it} - \text{black}_t$.

Thus, the regression model can be written as

$$p_{it} = \alpha_i + \sum_{j=1,2,3} \beta_j D_{jit} + \beta_4(w_i - w_t) + \beta_5(\text{play}_{it} - \text{play}_t) + \beta_6(\text{black}_{it} - \text{black}_t)$$

The main results can be found in table 11.1, where the columns refer to the following subsamples: Column 1 gives the results of the full sample of 39 clubs over the 16 years since the establishment of free agency in 1978, column 2 analyzes a subsample from the first 8 years, column 3 analyzes an equivalent one for the last 8 years, column 4 is based on a subsample of the 19 biggest clubs (measured in terms of average stadium capacity over the sample period), and column 5 reports the estimates for the 20 smallest clubs.

As can be seen from the table, the effects of both expenditure (relative wage bill) and turnover (number of players used) have the right sign and are highly significant, as we would expect.[4] Most interesting is the estimate of the coefficient of the variable measuring the proportion of appearances of black players. It is negative and statistically significant,

3 The league position is transformed into log odds to give greater weight to progression higher up in the league table.

4 Moreover, a conventional test for reverse causality, the Wu–Hausman test (using lagged wages and performance as instruments), indicates that there is no evidence of reverse causality. That is, it is not success that leads to high wages, but the other way around. Time dummies are also considered and do not change the conclusions.

**Table 11.1.** Determinants of Rank (Position) in the League Standings 1978–93

| | All Clubs | | | 19 Largest Clubs | 20 Smallest Clubs |
|---|---|---|---|---|---|
| | All Periods 1978–93 | First Half 1978–85 | Second Half 1986–93 | 1978–93 | 1978–93 |
| Relative wage bill | -0.535 | -0.691 | -0.557 | -0.632 | -0.368 |
| | (-0.169) | (-0.201) | (-0.236) | (-0.336) | (-0.172) |
| Number of players used | 2.067 | 1.704 | 2.323 | 2.257 | 1.885 |
| | (0.190) | (0.213) | (0.331) | (0.306) | (0.205) |
| Share of black players employed | -0.026 | -0.014 | -0.136 | -0.101 | -0.008 |
| | (0.011) | (0.010) | (0.049) | (0.052) | (0.007) |
| *p*-value for black player coefficient | 0.021 | 0.166 | 0.050 | 0.049 | 0.247 |
| Observations | 624 | 312 | 312 | 304 | 320 |

*Notes:* The dependent variable is Position (in log odds). Robust one-step standard errors are in parentheses.

implying that a club with a higher share of black players will systematically find itself in a better league position than its wage bill would indicate. Note that there is no additional financial cost associated with this improvement in sporting performance relative to discriminating teams. If the labor market for professional soccer players is efficient (and we have seen evidence consistent with this hypothesis), then these results represent evidence that black players were discriminated against and were therefore paid less than their talent justified. Equivalently, white players earned a premium over what their talent would justify.

There are two additional testable implications:

1. As we have seen above, one of the implications of the model is that if the share of black players in the overall population was sufficiently small, the premium discriminators would have to pay would be negligible. Therefore, we would expect the effect of discrimination to be smaller during 1978–85 (with a proportion of black professional soccer players of about 3%) than during 1986–93 (when the share of black players rises to 7.5%). Comparing columns 2 and 3 in the table, we immediately see that the size of the coefficient of the discrimination variable is indeed much larger during the later period.

2. Finally, the last two columns investigate the effect of discrimination of bigger clubs relative to smaller clubs. Soccer clubs with larger stadium capacities show a more pronounced discrimination effect than do smaller clubs. In other words, because the stakes are higher, discrimination is more costly in financial terms at higher levels of competition.

Although wage inflation means that the exact cost of discrimination also varies from year to year, Szymanski estimates that hiring no black players at all would have cost the average club an implicit penalty of 5% of its total wage bill, if it wanted to maintain any given position in the league, compared to a nondiscriminating team (one with a share of black players equal to the proportion of black players in the total player population). Given that the typical top club in 1993 spent around £5 million in wages, this percentage translates into a premium for discrimination of around £250,000 per year, relative to nondiscriminators.

Summing up, the pioneering contribution of this analysis is that it provides a way to test for taste-based discrimination in a labor market, without having to worry about the problem that plagued previous attempts to do so, the omitted variable bias. Though it is virtually impossible to observe all factors determining earnings, the crucial assumption that the market for professional players is competitive is much more likely to be satisfied here than in other settings, and indeed is a hypothesis supported by empirical evidence.[5] Under the assumption of labor market efficiency, the analysis provides strong empirical evidence that soccer clubs in England hiring a below-average proportion of black players during 1978–93 performed worse on the pitch than would seem warranted by their wage expenditure. It may be inferred from this evidence that clubs hiring a below-average share of black players belong to team owners with a taste for discrimination and pay a premium to satisfy this taste.

Two further points can be made before concluding this chapter.

5 Market efficiency is supported by the fact that wage expenditure alone can explain most of the variation in league position (see figure 11.1). However, if other variables potentially correlated with playing performance also turned out to be significant, this correlation would be a blow to the unbiasedness of a market-based test for racial discrimination. Szymanski also ran regressions with variables proxying for managerial performance, as well as the proportion of players developed within the youth teams of a club (as opposed to bought on the market) and the total number of players on a team. Though these variables tend to be correlated with performance, once the effect of wage expenditure is included in the specification, their effect is insignificant. This evidence, therefore, also supports the hypothesis of market efficiency.

1. Is discrimination by owners the only hypothesis that is supported by the evidence, or are other theories consistent with it as well? Although the evidence is consistent with the hypothesis of discriminating owners, others have suggested the hypothesis of fan discrimination, with owners simply responding rationally to fan tastes. Here Szymanski reports regressions of both attendance at league matches and annual revenues on league performance, the proportion of black players, and other relevant variables (such as ticket prices and success in cup competitions). It turns out that the share of black players is not a significant explanatory variable in either the revenues or the attendance regression, something which provides some evidence against the hypothesis of fan-based discrimination.
2. How has the situation evolved since the end of the sample period 1978–93, say over the *following* 16-year period, 1993–2008? Specifically, is there any evidence that competition alleviates this form of racial discrimination over time?

Kenneth J. Arrow (1973) contends that if the market in question were efficient then nondiscriminating profit-maximizers, taking advantage of the underpriced asset, would be able to drive the discriminating firms out of the market over time. In other words, in a competitive market, this type of discrimination would be competed away since nondiscriminatory clubs would employ black players, given their lower cost (at any given level of talent). Nondiscriminating clubs therefore would make higher profits than discriminating soccer clubs and would eventually drive the latter out of the market. In the process, black players' wages would be bid up until the wage would be independent of race.

In the sample period examined by Szymanski, the market for corporate control of English professional soccer clubs was still very limited and generally management policies were extremely conservative. However, since the early 1980s, a market for corporate control has emerged. In many cases, clubs have become insolvent and new owners have bailed out clubs and introduced new ideas. The first major club to list its shares on the stock market was Tottenham Hotspur in 1983, followed by Manchester United in 1991. In the 1990s, there was a flurry of clubs listing on the stock exchange. In 1996–97, for example, another 18 English clubs raised capital on the stock market in one form or another, and whereas most of these have now delisted, this process also led to significant turnover in ownership. More generally, the growth in the profile of soccer in the 1990s made ownership more sought after, not least with an increasing number of foreign owners. All these phenomena have made English soccer both more competitive and more open to new ideas.

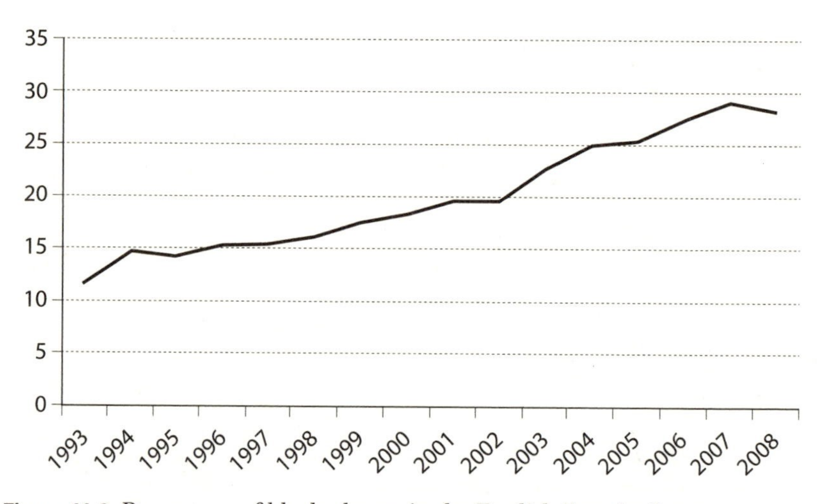


**Figure 11.2.** Percentage of black players in the English Premier League 1993–2008.

Besides an even more competitive labor market, the shares of black players and black player appearances have continued to increase relentlessly during the period 1993–2008. In the same sample of 39 teams as in the previous 16-year period, the share of black players has basically tripled from around 10% in 1993 to close to 30% in 2008 (see figure 11.2). Furthermore, in 1993 there were 2,033 appearances by black players; in 2008, there are 4,512 line-up appearances (F), plus 1,307 as substitutes (Sub) (see figure 11.3).

The theory suggests that as a result, the wage premium for white players should shrink over the years, and perhaps even disappear. The evidence in table 11.2 answers this question. As can be seen, the regression estimates confirm this hypothesis. The coefficients have the same signs as in the previous period, but the effect of the race variable has disappeared, to the point that the coefficient is statistically insignificant and identical to zero.

The recent review of the research literature on discrimination by Lang and Lehmann (2012) shows that in recent years the pioneering approach of the market-based test of discrimination has been quite fruitful. This review also shows that sports settings have been particularly valuable; a number of important contributions have been obtained in these settings.[6]

6 Prominent academic studies include Price and Wolfers (2010) and Parsons et al. (2011).

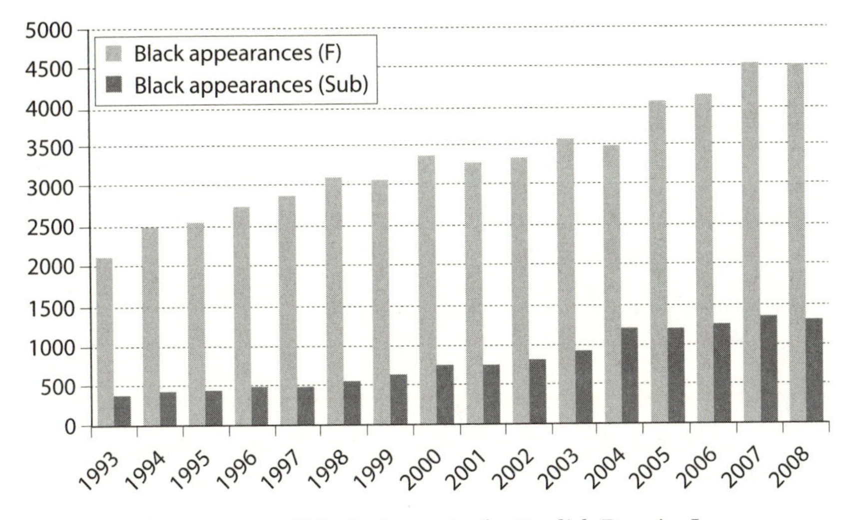


**Figure 11.3.** Appearances of black players in the English Premier League 1993–2008.

**Table 11.2.** Determinants of Rank (Position) in the League Standings 1993–2008

| | All Clubs | | | 19 Largest Clubs | 20 Smallest Clubs |
|---|---|---|---|---|---|
| | All Periods 1993–2008 | First Half 1993–2000 | Second Half 2001–8 | 1993–200 | 2001–8 |
| Relative wage bill | -0.735 | -0.681 | -0.797 | -0.868 | -0.518 |
| | (-0.144) | (-0.181) | (-0.217) | (-0.156) | (-0.142) |
| Number of players used | 2.292 | 1.984 | 2.673 | 2.272 | 2.185 |
| | (0.172) | (0.288) | (0.301) | (0.286) | (0.227) |
| Share of black players employed | -0.008 | -0.004 | -0.010 | -0.021 | -0.001 |
| | (0.010) | (0.010) | (0.011) | (0.024) | (0.006) |
| *p*-value for black player coefficient | 0.821 | 0.898 | 0.750 | 0.779 | 0.947 |
| Observations | 624 | 312 | 312 | 304 | 320 |

*Notes:* The dependent variable is position (in log odds). Robust one-step standard errors are in parentheses.

But besides studying *conscious* forms of discrimination like statistical-based discrimination and taste-based discrimination, it is also possible to study *unconscious* or *implicit* forms of discrimination for the first time, also using the same sports setting. *Implicit discrimination* is a form of discrimination based on the unconscious mental association between members of a social group and a given negative attribute. It has received a lot of attention from psychologists, whereas economists have tended to focus on conscious forms of discrimination.

Gallo, Grund, and Reade (2013) present the first empirical study on implicit discrimination in a natural setting using, again, a large data set on soccer matches in the English Premier League (EPL). The data set contains 1,050,411 in-match events, and the authors investigate discrimination by referees against players belonging to specific social groups, in particular against players of *oppositional identity* to the referees.

The concept of oppositional identity comes from Akerlof and Kranton (2010), who argue that the strongest determinant of the formation of an oppositional identity is a difference in socioeconomic opportunities. All the referees in the EPL sample were born in the United Kingdom and have been living in the United Kingdom throughout their lives. Thus, oppositional identity players are defined as players who are foreign, nonwhite, and from the same background as the most sizable minorities in the United Kingdom. The empirical evidence shows that white referees award significantly *more* yellow cards against nonwhite players of oppositional identity, controlling for several player, team, referee, and match characteristics. Moreover, there is no significant difference in the probability of receiving a booking for other social groups defined along purely racial (i.e., Asians/blacks/whites), nationality (i.e., UK/foreign), or linguistic criteria.

Furthermore, according to Bertrand et al. (2005), there are two ways to empirically differentiate implicit discrimination from other forms of conscious discrimination. First, the level of implicit discrimination should be positively related to the level of ambiguity because an ambiguous situation increases the cognitive load on the brain, which reduces this load by resorting to the automatized response schemes that lead to implicit discrimination. The second suggestion is that implicit discrimination should be negatively related to the amount of time the discriminator has to make his or her decision. Time pressure demands an accelerated mental processing, which the brain achieves by falling back on automatized responses. Gallo, Grund, and Reade (2013) find that in fact discrimination (1) increases according to how rushed the referee is before making a decision (in particular, in parts of the field where players are rushing to put the ball back in play to orchestrate a counterattack) and (2) increases according to the level of ambiguity of the decision (that is, it is lower for unambiguous yellow cards than for ambiguous ones, e.g., bookings for excessive celebrations or referee abuse, and for typically unambiguous red cards).

*

The empirical evidence in this chapter suggests that Nelson Mandela's intuition was correct: "Sport has the power to change the world. It has the power to inspire, the power to unite people that little else has." If we look closely, I believe we will find many examples where sports critically help break down racial barriers, perhaps not as clean from a statistical perspective as the evidence from the EPL presented in this chapter, but intuitively quite convincing.

And to do this job, *winning* helps. For example, much as seeing the Springboks win the Rugby World Cup in 1995 contributed to unite South African people, winning the 1998 soccer World Cup helped France to be understood, finally, as a multiracial country.

As Relaño (2010) indicates, for a long time, players of various races, particularly from the Maghreb or south of the Sahara, had appeared on the national team of France. In the mid-1990s, they became the majority of players. This situation came to irritate the far-right leader Jean-Marie Le Pen, president of the National Front. In June 1996, at a rally of his political party in Saint-Gilles, he said, "It is artificial to make foreigners come and then baptize them as the French team." Besides, he complained, these players typically did not sing the *Marseillaise* before the games: "I do not know if they do not want to or it is because they do not know it." He promised to review the status of these players if and when he took office.

Players on the national team felt quite uneasy and upset. They even called to vote against Le Pen in the general elections. Although the French players came from a variety of origins, the fact is that all were born in France or its colonies, with the exception of Marcel Desailly, born in Ghana and nationalized French.

France had a good national team. They had reached the semifinals of the 1996 European Football Championship in England (where they drew 0–0 with the Czech Republic but lost in the penalty shoot-out despite kicking first) when the 1998 World Cup was about to begin on home soil in France. To the horror of Le Pen, there were only eight players selected for the national team whom he could consider "pure," white children of French father and French mother. The rest originated from Arab, Caribbean, South American, and African countries, and even one from New Caledonia in the South Pacific (Christian Karembeu). The team was led by Zinedine Zidane, born in the neighborhood of La Castellane in Marseille of Algerian parents. Zidane himself came to speak out on the dispute between Le Pen and the national team: "I am French. My father is Algerian. I am proud to be French, and I am proud that my father is Algerian."

France played extremely well during the whole tournament and reached the final against Brazil. In the final, France was leading 2–0 at halftime, both goals from Zidane. Despite a push in the second half, Brazil did not score, and Emmanuel Petit, white and blond, one of those whom Le Pen would approve, scored the third goal in the last minute of the game. It was the first time that France had won the World Cup. A burst of unprecedented dimensions exploded across France, the biggest ever seen in the country, with millions of people just on the Champs Elysées. President Chirac called it "this multicolored winning France," explicitly recognizing the country as a multiracial community.

# ACKNOWLEDGMENTS

FOUR CHAPTERS ARE BASED ON PREVIOUSLY PUBLISHED ARTICLES OF MINE, AS a single author (chapter 1) and with coauthors (chapters 2, 5, and 7). I want to thank my coauthors in these chapters: Oscar Volij, Luis Garicano, Canice Prendergast, and Jose Apesteguia. It was a pleasure working with and learning from them. Two more chapters are based on articles by Karen Croxson and J. James Reade (chapter 6), and Stefan Szymanski (chapter 11), respectively. I thank them for their generosity reviewing these chapters. These six chapters based on previous research differ, sometimes markedly, from the original articles. Nevertheless, there are several places where tables, figures, and text have been borrowed from the published papers.

The rest of the chapters involved new research and have not been published previously. Chapter 4 is jointly written with Antonio Olivero, Sven Bestmann, Jose Florensa Vila, and Jose Apesteguia. We are grateful to the Hospital Nacional de Parapléjicos de Toledo (Spain) and University College London for allowing us to use their fMRI facilities and for their hospitality. Chapter 8 is joint with Luis Garicano, and we received valuable research assistance from Michael Suh. For chapter 10, Martín Guzmán first told me about the penalty shoot-out experiment that was run in Argentina in the 1988–89 season, and Oscar Volij and Jonathan Volij collected the data on the order that teams followed in the shoot-outs from the archives of the Biblioteca Nacional de la República Argentina in Buenos Aires.

A number of other friends and colleagues have helped me with their insights, suggestions, and advice in specific chapters and, more generally, with conversations over the years on different topics and areas studied in this book. These include Cyrus Amir-Mokri, María Azketa, Jordi Blanes i Vidal, Seamus Brady, Colin Camerer, Juan Carrillo, Pedro Dal Bó, Sir Howard Davies, David De Meza, Jeff Ely, Luis Garicano, José Ángel Iribar, Tony Lancaster, Hans Leitert, Eugenio Miravete, Gerard Padró i Miguel, Patxo Palacios, Canice Prendergast, Yona Rubinstein, Ana Saracho, Catherine Thomas, Josu Urrutia, and Oscar Volij.

I am particularly grateful to Gary S. Becker and Kevin Murphy for the enormous influence they have had shaping my brain as an economist. Most of what I know about the economic approach to human behavior I learned from them. Together with Oscar Volij, they are the greatest economists I know and, more importantly, good friends.

Karen Croxson, J. James Reade, Bradley Ruffle, Jakob Schneebacher, Catherine Thomas, and Jan-Joost van den Bogert provided helpful editorial assistance that substantially improved the manuscript. I also thank John Carlin, Enric González, Alfredo Relaño, Clark Miller, and James Miller for allowing me to adapt and present some of the stories and anecdotes they first reported.

I am deeply grateful to Seth Ditchik for first suggesting the idea for this book and for his patience. Richard Baggaley was also extremely useful at the initial stages of the project and, importantly, he suggested along with David De Meza (but independently) the title of the book. Beth Clevenger, Natalie Baan, and Paula Bérard provided terrific help and assistance at every step of the publishing process.

Finally, I want to thank my wife Ana and my children Ander and Julene for their love and for giving me a life I never dreamt of. My mother, my late father, and my brothers deserve the same thanks. This book would have never been possible without the love, support, and environment they have always provided.

# REFERENCES


Akerlof, George A. "A Theory of Social Custom, of Which Unemployment May Be One Consequence." *Quarterly Journal of Economics* 94, no. 4 (1980): 749–75.

Akerlof, George A., and Rachel E. Kranton. *Identity Economics: How Our Identities Shape Our Work, Wages, and Well-Being*. Princeton, NJ: Princeton University Press, 2010.

Allouche, Jean-Paul, and Jeffrey Shallit. "The Ubiquitous Prouhet–Thue–Morse Sequence." In *Sequences and Their Applications: Proceedings of SETA '98,* edited by C. Ding, T. Helleseth, and H. Niederreiter, 1–16. London: Springer-Verlag, 1999.

Alonso, Ricardo, Isabelle Brocas, and Juan D. Carrillo. "Resource Allocation in the Brain." *Review of Economic Studies* (forthcoming).

Anderson, Chris, and David Sally. *The Numbers Game*. London: Penguin Group, 2013.

Anderson, Kathryn, Richard Burkhauer, and Jennie Raymond. "The Effect of Creaming on Placement Rates under the Job Training Partnership Act." *Industrial and Labor Relations Review* 46, no. 4 (1993): 613–24.

Andersson, Jesper L. R., Chloe Hutton, John Ashburner, Robert Turner, and Karl Friston. "Modeling Geometric Deformations in EPI Time Series." *Neuroimage* 13 (2001): 903–19.

Apesteguia, Jose, and Ignacio Palacios-Huerta. "Psychological Pressure in Competitive Environments: Evidence from a Randomized Natural Experiment." *American Economic Review* 100, no. 5 (2010): 2548–64.

Arellano, Manuel, and Raquel Carrasco. "Binary Choice Panel Data Models with Predetermined Variables." *Journal of Econometrics* 115, no. 1 (2003), 125–57.

Arellano, Manuel, and Bo Honoré. "Panel Data Models: Some Recent Developments." In *Handbook of Econometrics,* edited by James E. Heckman and Edward Leamer, vol. 5, chap. 53, 3229–96. Amsterdam: North-Holland, 2001.

Ariely, Dan, Uri Gneezy, George Loewenstein, and Nina Mazar. "Large Stakes and Big Mistakes." *Review of Economic Studies* 76, no. 2 (2009): 451–69.

Arrow, Kenneth J. "The Theory of Discrimination." In *Discrimination in Labor Markets,* edited by O. Ashenfelter and A. Rees. Princeton, NJ: Princeton University Press, 1973.

Aumann, Robert J. "Game Theory." In *The New Palgrave: A Dictionary of Economics,* edited by J. Eatwell, M. Milgate, and P. Newman, vol. 2, 460–82. London: MacMillan Press, 1987.

Bachelier, Louis. "Théorie de la Spéculation." In *The Random Character of Stock Market Prices,* edited by P. Cootner, 17–78. Cambridge, MA: MIT Press, 1900.

Baker, George. "Incentive Contracts and Performance Measurement." *Journal of Political Economy* 100, no. 3 (1992): 598–614.

Bar-Eli, Michael, Ofer H. Azar, Ilana Ritov, Yael Keidar-Levin, and Galit Schein. "Action Bias among Elite Soccer Goalkeepers: The Case of Penalty Kicks." *Journal of Economic Psychology* 28, no. 5 (2007): 606–21.

Bar-Hillel, Maya, and Willem A. Wagenaar. "The Perception of Randomness." *Advances in Applied Mathematics* 12, no. 4 (1991): 428–54.

Becker, Gary S. *Accounting for Tastes*. Cambridge, MA: Harvard University Press, 1996.

——. *The Economic Approach to Human Behavior*. Chicago: University of Chicago Press, 1976.

——. *The Economics of Discrimination*. Chicago: University of Chicago Press, 1957.

Becker, Gary S., and Kevin M. Murphy. *Social Economics: Market Behavior in a Social Environment*. Cambridge, MA: Harvard University Press, 2000.

Becker, Gary S., and Yona Rubinstein. "Fear and the Response to Terrorism: An Economic Analysis." LSE manuscript, 2013.

Beilock, Sian. *Choke: What the Secrets of the Brain Reveal about Getting It Right When You Have To*. New York: Free Press, 2010.

Belot, Michèle, Vincent P. Crawford, and Cecilia Heyes. "Players of Matching Pennies Automatically Imitate Opponents' Gestures against Strong Incentives." *Proceedings of the National Academy of Sciences*, 110, no. 5 (February 19, 2013).

Bernheim, Douglas B. "A Theory of Conformity." *Journal of Political Economy* 102, no. 5 (1994): 841–77.

Bertrand, Marianne, Dolly Chugh, and Shendil Mullainathan. "Implicit Discrimination." *American Economic Review* 95, no. 2 (2005): 94–98.

Best, George. *Scoring at Half-Time*. London: Ebury Press, 2004.

Bhaskar, V. "Rational Adversaries? Evidence from Randomized Trials in One Day Cricket." *Economic Journal* 119, no. 535 (February 2009): 1–23.

Boffrey, Philip M. "The Next Frontier Is Inside Your Brain." *New York Times*, February 23, 2013.

Bowen, Huw V. "Investment and Empire in the Later Eighteenth Century: East India Stockholding, 1756–1791." *The Economic History Review* 42, no. 2 (1989): 186–206.

Brams, Steven J., and Alan D. Taylor. *The Win-Win Solution: Guaranteeing Fair Shares to Everybody*. New York: W. W. Norton, 1999.

Brocas, Isabelle, and Juan D. Carrillo. "The Brain as a Hierarchical Organization." *American Economic Review* 98, no. 4 (2008): 1312–46.

Brown, James, and Robert Rosenthal. "Testing the Minimax Hypothesis: A Reexamination of O'Neill's Experiment." *Econometrica* 58, no. 5 (1990): 1065–81.

Bull, Clive, Andrew Schotter, and Keith Weigelt, "Tournaments and Piece Rates: An Experimental Study." *Journal of Political Economy* 95, no. 1 (1987): 1–31.

Cabral, Luis. "Increasing Dominance with No Efficiency Effect." *Journal of Economic Theory* 102, no. 2 (2002): 471–79.

——. "R&D Competition When Firms Choose Variance." *Journal of Economics & Management Strategy* 12, no. 1 (2003): 139–50.

Camerer, Colin F. *Behavioral Game Theory*. Princeton, NJ: Princeton University Press, 2003.

——. "Individual Decision Making." In *Handbook of Experimental Economics,* edited by J. H. Hagel and A. E. Roth, 587–703. Princeton, NJ: Princeton University Press, 1995.

——. "Neuroeconomics: Opening the Gray Box." *Neuron* 60, no. 3 (November 6, 2008): 416–19.

——. "The Promise and Success of Lab-Field Generalizability in Experimental Economics: A Critical Reply to Levitt and List." Working paper, California Institute of Technology, Pasadena, CA, 2011). http://ssrn.com/abstract=1977749.

Camerer, Colin F., George Loewenstein, and Drazen Prelec. "Neuroeconomics: How Neuroscience Can Inform Economics." *Journal of Economic Literature* 43, no. 1 (2005): 9–64.

Caplin, Andrew. "Fear as a Policy Instrument." In *Time and Decision,* edited by George Loewenstein and Daniel Read. New York: Russell Sage, 2003.

Caplin, Andrew, and John Leahy. "Psychological Expected Utility Theory and Anticipatory Feelings." *Quarterly Journal of Economics* 116, no. 1 (2001): 55–80.

——. "The Supply of Information by a Concerned Expert." *Economic Journal* 114 (2004), 487–505.

Carlin, John. "Los Dioses Pisan el Césped." *El País* (Madrid, Spain), July 12, 2009.

——. *Playing the Enemy: Nelson Mandela and the Game That Made a Nation.* London: The Penguin Press, 2008.

Che, Yeon-Koo, and Terry Hendershott. "How to Divide the Possession of a Football?" *Economics Letters* 99, no. 3 (2008): 561–65.

Chiappori, Pierre-Andre, Steven D. Levitt, and Timothy Groseclose. "Testing Mixed Strategy Equilibrium When Players Are Heterogeneous: The Case of Penalty Kicks." *American Economic Review* 92, no. 4 (2002): 1138–51.

Chowdhury, Subhasish M., and Oliver Gürtler. "Sabotage in Contests: A Survey." University of East Anglia, Mimeo, Norwich, 2013.

Cook, Richard, Geoffrey Bird, Gabriele Lünser, Steffen Huck, and Cecilia Heyes. "Automatic Imitation in a Strategic Context: Players of Rock–Paper–Scissors Imitate Opponents' Gestures." *Proceedings of the Royal Society B: Biological Sciences*, 1729 (2011): 780–86.

Cooper, Joshua, and Aaron Dutle. "Greedy Galois Games." *The American Mathematical Monthly* 120, no. 5 (May 2013): 441–51.

Courty, Pascal, and Gerald Marschke. "An Empirical Investigation of Gaming Responses to Explicit Performance Incentives." *Journal of Labor Economics* 22, no. 1 (2004): 23–56.

Cragg, Michael. "Performance Incentives in the Public Sector: Evidence from the Job Training Partnership Act." *Journal of Law, Economics and Organization* 13, no. 1 (1997): 147–68.

Croxson, Karen, and J. James Reade. "Exchange vs. Dealers: A High-Frequency Analysis of In-Play Betting Prices." Discussion Paper 11–19, Birmingham, UK: Department of Economics, University of Birmingham, 2011.

——. "Information and Efficiency: Goal Arrival in Soccer Betting." *The Economic Journal.* Published electronically June 3, 2013. doi:10.1111/ecoj.12033.

Daniels, Christine, Karsten Witt, S. Wolff, Olav Jansen, and Günther Deuschl. "Rate Dependency of the Human Cortical Network Subserving Executive Functions During Generation of Random Number Series: A Functional Magnetic Resonance Imaging Study." *Neuroscience Letters* 345 (2003): 25–28.

Dawson, Peter, and Stephen Dobson. "The Influence of Social Pressure and Nationality on Individual Decisions: Evidence from the Behaviour of Referees." *Journal of Economic Psychology* 31, no. 2 (2010): 181–91.

DellaVigna, Stefano. "Psychology and Economics: Evidence from the Field." *Journal of Economic Literature* 47, no. 2 (2009): 315–72.

Deloitte. *Deloitte Annual Review of Football Finance 2013*. London: Deloitte, 2013.

Dixit, Avinash, and Robert Pindyck. *Investment under Uncertainty*. Princeton, NJ: Princeton University Press, 1994.

Dobson, Stephen, and John Goddard. *The Economics of Football*. Cambridge, UK: Cambridge University Press, 2011.

Doyle, Arthur Conan. "Silver Blaze." In *The Memoirs of Sherlock Holmes*. London: George Newnes, 1892.

Drago, Robert, and Gerald T. Garvey. "Incentives for Helping on the Job: Theory and Evidence." *Journal of Labor Economics* 16, no. 1 (1998): 1–25.

Driver, P. M., and N. Humphries. *Protean Behavior: The Biology of Unpredictability*. Oxford, UK: Oxford University Press, 1988.

Dwyre, Bill. "FIFA Approves Scoring Changes." *Los Angeles Times*, Dec. 17, 1993.

Edgeworth, Francis Y. *Mathematical Psychics: An Essay on the Application of Mathematics to the Moral Sciences*. London, UK: C. Kegan Paul & Co., 1881.

Ehrenberg, Ronald G., and Michael L. Bognanno. "Do Tournaments Have Incentive Effects?" *Journal of Political Economy* 98, no. 6 (1990): 1307–24.

Ericsson, K. Anders, Neil Charness, Paul J. Feltovich, and Robert R. Hoffman, eds. *The Cambridge Handbook of Expertise and Expert Performance*. Cambridge, MA: Cambridge University Press, 2006.

Falk, A., and James Heckman. "Lab Experiments Are a Major Source of Knowledge in the Social Sciences." *Science* 326, no. 5952 (2009): 535–38.

Fama, Eugene F. "The Behavior of Stock Market Prices." *Journal of Business* 38, no. 1 (1965): 34–105.

——. "Efficient Capital Markets: A Review of Theory and Empirical Work." *Journal of Finance* 25, no. 2 (1970): 383–417.

——. "Market Efficiency, Long-Term Returns and Behavioral Finance." *Journal of Financial Economics* 49, no. 3 (1998): 283–306.

Fédération Internationale de Football Association (FIFA). *Laws of the Game 2012/2013*. Zurich, Switzerland: Fédération Internationale de Football Association, 2012.

Friston, Karl J., P. Fletcher, O. Josephs, Andrew P. Holmes, M. D. Rugg, and R. Turner. "Event-Related fMRI: Characterizing Differential Responses." *Neuroimage* 7 (1998): 30–40.

Friston, Karl J., Andrew P. Holmes, J. B. Poline, P. J. Grasby, S. C. Williams, R. S. Frackowiak, and R. Turner. "Analysis of fMRI Time-Series Revisited." *Neuroimage* 2 (1995): 45–53.

Galeano, Eduardo. "Pobre Mi Madre Querida." In *El Fútbol a Sol y a Sombra*. Madrid: Ediciones Siglo XXI, 1995.

Gallo, Edoardo, Thomas Grund, and J. James Reade. "Punishing the Foreigner: Implicit Discrimination in the Premier League Based on Oppositional Identity." *Oxford Bulletin of Economics and Statistics* 75, no. 1 (2013): 136–56.

Game Theory Society. www.gametheorysociety.org. Accessed July 27, 2006.

Garicano, Luis, Ignacio Palacios-Huerta, and Canice Prendergast. "Favoristism under Social Pressure." NBER Working Paper no. 8376, National Bureau of Economic Research, Cambridge, MA, 2001.

——. "Favoritism under Social Pressure." *Review of Economics and Statistics* 87, no. 2 (2005): 208–16.

Genakos, Christos, and Mario Pagliero. "Interim Rank, Risk Taking, and Performance in Dynamic Tournaments." *Journal of Political Economy* 120, no. 4 (2012): 782–813.

Gibbons, Jean D., and Subhabrata Chakraborti. *Nonparametric Statistical Inference*. New York: Marcel Dekker, 1992.

Gibbons, Robert. "Incentives in Organizations." *Journal of Economic Perspectives* 12, no. 4 (1998): 115–32.

Glimcher, Paul W., Colin F. Camerer, Russell A. Poldrack, and Ernst Fehr. *Neuroeconomics. Decision Making and the Brain*. Oxford, UK: Academic Press, 2009.

González, Enric. "El Balón y la Bandera." *El País* (Madrid), May 31, 2008.

Gonzalez-Díaz, Julio, and Ignacio Palacios-Huerta. "Cognitive Performance in Dynamic Tournaments." London School of Economics, Mimeo, 2012.

Green, Jerry R., and Nancy L. Stokey. "A Comparison of Tournaments and Contracts." *Journal of Political Economy* 91, no. 3 (1983): 349–64.

Grether, David M., and Charles R. Plott. "The Effects of Market Practices in Oligopolistic Markets: An Experimental Examination of the Ethyl Case." *Economic Inquiry* 22, no. 4 (1984): 479–507.

Gul, Faruk, and Wolfgang Pesendorfer. "The Case for Mindless Economics." In *The Foundations of Positive and Normative Economics,* edited by Andrew Caplin and Andrew Schotter, New York: Oxford University Press, 2008.

Hahn, Robert W., and Paul C. Tetlock. *Information Markets: A New Way of Making Decisions*. Cambridge, MA: AEI Press, 2006.

Hampton, Alan N., Peter Bossaerts, and John P. O'Doherty. "Neural Correlates of Mentalizing-Related Computations during Strategic Interactions in Humans." *Proceedings of the National Academy of Sciences* 105, no. 18 (May 6, 2008): 6741–46.

Harrison, Glenn W., and John A. List. "Field Experiments." *Journal of Economic Literature* 42, no. 4 (2004): 1009–55.

Heckman, James J. 2008. "Schools, Skills and Synapses." *Economic Inquiry* 46, no. 3, (2008): 289–324.

Hirschman, Albert. *The Passions and the Interests: Political Arguments for Capitalism before Its Triumph*. Princeton, NJ: Princeton University Press, 1977.

Holmstrom, Bengt, and Paul Milgrom. "The Firm as an Incentive System." *American Economic Review* 84, no. 4 (1994): 972–91.

——. "Multitask Principal-Agent Analyses: Incentive Contracts, Asset Ownership, and Job Design." *Journal of Law, Economics and Organization* 7 (1991): 24–52.

Hong, James T., and Charles R. Plott. "Rate Filing Policies for Inland Water Transportation: An Experimental Approach." *Bell Journal of Economics* 13, no. 1 (1982): 1–19.

Hume, David. *A Treatise of Human Nature,* 2nd ed. Oxford, UK: Clarendon Press, 1739.

Hvide, Hans K. "Tournament Rewards and Risk Taking." *Journal of Labor Economics* 20, no. 4 (2002): 877–98.

Hvide, Hans K., and Eirik G. Kristiansen. "Risk Taking in Selection Contests." *Games and Economic Behavior* 42, no. 1 (2003): 172–79.

Ischebeck, Anja, Stefan Heim, Christian Siedentopf, Laura Zamarian, Michael Schocke, Christian Kremser, Karl Egger, Hans Strenge, Filip Scheperjans, and Margarete Delazer. "Are Numbers Special? Comparing the Generation of Verbal Materials from Ordered Categories (Months) to Numbers and Other Categories (Animals) in an fMRI Study." *Human Brain Mapping* 29 (2008): 894–909.

Itoh, Hideshi. "Cooperation in Hierarchical Organizations: An Incentive Perspective." *Journal of Law, Economics, and Organization* 8, no. 2 (1992): 321–45.

——. "Incentives to Help in Multi-Agent Situations." *Econometrica* 59, no. 3 (1991): 611–36.

Jacob, Brian A., and Steven D. Levitt. "Rotten Apples: An Investigation of the Prevalence and Predictors of Teacher Cheating." *Quarterly Journal of Economics* 118, no. 3 (2003): 843–77.

Jahanshahi, Marjan, Paolo Profice, Richard G. Brown, Mike C. Ridding, Georg Dirnberger, and John C. Rothwell. "The Effects of Transcranial Magnetic Stimulation over the Dorsolateral Prefrontal Cortex on Suppression of Habitual Counting During Random Number Generation." *Brain* 121 (1998): 1533–44.

Jewell, R. Todd, Rob Simmons, and Stefan Szymanski. "Bad for Business? The Effect of Hooliganism on English Professional Soccer Clubs." University of Michigan, Mimeo, Ann Arbor, 2013.

Kadota, Hiroshi, Hirofumi Sekiguchi, Shigeki Takeuchi, Makoto Miyazaki, Yutaka Kohno, and Yasoichi Nakajima. "The Role of the Dorsolateral Prefrontal Cortex in the Inhibition of Stereotyped Responses." *Experimental Brain Research* 203 (2010): 593–600.

Kahn, Lawrence M. "The Sports Business as a Labor Market Laboratory." *Journal of Economic Perspectives* 14, no. 3 (2000): 75–94.

Kahneman, Daniel, and Amos Tversky. "Prospect Theory: An Analysis of Decision under Risk." *Econometrica* 47, no. 2 (1979): 263–92.

Kerr, Steven. "On the Folly of Rewarding *A*, While Hoping for *B*." *Academy of Management Journal* 18 (1975): 769–83.

Köszegi, Botond, and Matthew Rabin. "A Model of Reference Dependent Preferences." *Quarterly Journal of Economics* 121, no. 4 (2006): 1133–66.

Koudijs, Peter. "The Boats That Did Not Sail: Asset Price Volatility and Market Efficiency in a Natural Experiment." Working Paper No. 18831, National Bureau of Economic Research, Cambridge, MA, 2013.

Kreps, David M. *Game Theory and Economic Modelling*. Oxford, UK: Oxford University Press, 1991.

Kuper, Simon. *The Football Men*. London: Simon & Schuster, 2011.

Kuper, Simon, and Stefan Szymanski. *Soccernomics*. New York: Nation Books, 2012.

Lang, Kevin, and Jee-Yeon K. Lehmann. "Racial Discrimination in the Labor Market: Theory and Empirics." *Journal of Economic Literature* 50, no. 4 (2012): 959–1006.

Lazear, Edward P. "Pay Equality and Industrial Politics." *Journal of Political Economy* 97, no. 3 (1989): 561–80.

Lazear, Edward P., and Sherwin H. Rosen. "Rank-Order Tournaments as Optimum Labor Contracts." *Journal of Political Economy* 89, no. 5 (1981): 841–64.

Levine, Lionel, and Katherine E. Stange. "How to Make the Most of a Shared Meal: Plan the Last Bite First." *The American Mathematical Monthly* 119, no. 7 (2012): 550–65.

Levitt, Steve D., John A. List, and David Reiley. "What Happens in the Field Stays in the Field: Professionals Do Not Play Minimax in Laboratory Experiments." *Econometrica* 78, no. 4 (2010): 1413–34.

Loewenstein, George. "Anticipation and the Value of Delayed Consumption." *Economic Journal* 97, no. 387 (1987): 666–84.

Macrae, Norman. *John Von Neumann: The Scientific Genius Who Pioneered the Modern Computer, Game Theory, Nuclear Deterrence, and Much More,* 1st ed. New York: Pantheon Books, 1992.

Mankiw, N. Gregory. "What Stock to Buy? Hey, Mom, Don't Ask Me." *New York Times,* May 18, 2013.

Manski, Charles F. *Identification Problems in the Social Sciences*. Cambridge, MA: Harvard University Press, 1995.

———. "Interpreting the Predictions of Prediction Markets." *Economic Letters* 91, no. 3 (2006): 425–29.

*Marca. Guía de la Liga 2013*. Madrid: Recoletos Grupo de Comunicación, 2012.

Martin, Christopher Flynn, Rahul Bhui, Peter Bossaerts, Tetsuro Matsuzawa, and Colin Camerer. "Experienced Chimpanzees Behave More Game-Theoretically than Humans in Simple Competitive Interactions." Caltech, Mimeo, Pasadena, 2013.

Maskin, Eric. "Can Neural Data Improve Economics?" *Science* 321 (2008): 1788–89.

Mertel, Manuel. "Under Pressure: Evidence from Repeated Actions by Professional Sportsmen." Master's thesis, Centro de Estudios Monetarios y Financieros, Madrid, 2011.

Miller, Clark. *He Always Puts It to the Right: A Concise History of the Penalty Kick*. London: Victor Gollancz, 1998.

Miller, Geoffrey F. "Protean Primates: The Evolution of Adaptive Unpredictability in Competition and Courtship." In *Machiavellian Intelligence II: Extensions and Evaluations,* edited by A. Whiten and R. Byrne, 2nd ed. Cambridge, UK: Cambridge University Press, 1997.

Mookherjee, Dilip, and Barry Sopher. "Learning Behavior in an Experimental Matching Pennies Game." *Games and Economic Behavior* 7, no. 1 (1994): 62–91.

Morse, Marston. "Recurrent Geodesics on a Surface of Negative Curvature." *Transactions of the American Mathematical Society* 22 (1921): 84–100.

Moskowitz, Tobias J., and L. Jon Wertheim. *Scorecasting. The Hidden Influences behind How Sports Are Played and Games Are Won*. New York: Crown Archetype, 2011.

Nalebuff, Barry J., and Joseph E. Stiglitz. "Prizes and Incentives: Towards a General Theory of Compensation and Competition." *Bell Journal of Economics* 14, no. 1 (1983): 21–43.

Nash, John F., Jr. "Equilibrium Points in *N*-person Games." *Proceedings of the National Academy of Sciences* 36, no. 1 (1950): 48–49.

——. "Non-Cooperative Games." *Annals of Mathematics* 54, no. 2 (1951): 286–95.

Neuringer, Allen J. "Can People Behave 'Randomly'?: The Role of Feedback." *Journal of Experimental Psychology: General* 115 (1986): 62–75.

——. "Operant Variability: Evidence, Function, and Theory." *Psychonomic Bulletin and Review* 9 (2002): 672–705.

O'Neill, Barry. "Comments on Brown and Rosenthal's Reexamination." *Econometrica* 59, no. 2 (1991): 503–7.

——. "Nonmetric Tests of the Minimax Theory of Two-Person Zerosum Games." *Proceedings of the National Academy of Sciences* 84 (1987): 2106–9.

Osborne, Martin, and Ariel Rubinstein. *A Course on Game Theory*. Cambridge, MA: MIT Press, 1994.

Ozgit, Alper. "Auction Versus Dealer Markets in Online Betting." Working Paper, Department of Economics, UCLA, Los Angeles, 2005.

Palacios-Huerta, Ignacio. "Professionals Play Minimax." *Review of Economic Studies* 70, no. 2 (2003): 395–415.

Palacios-Huerta, Ignacio, Antonio Olivero, Sven Bestmann, Jose Florensa Vila, and Jose Apesteguia. "Mapping Minimax in the Brain: Supplementary Material." London School of Economics, Mimeo, 2013.

Palacios-Huerta, Ignacio, and Oscar Volij. "Experientia Docet: Professionals Play Minimax in Laboratory Experiments." *Econometrica* 76, no. 1 (2008): 71–115.

Parsons, Christopher A., Johan Sulaeman, Michael C. Yates, and Daniel S. Hamermesh. "Strike Three: Discrimination, Incentives and Evaluation." *American Economic Review* 101, no. 4 (2011): 1410–35.

Pettersson-Lidbom, Per, and Mikael Priks. "Behavior under Social Pressure: Empty Italian Stadiums and Referee Bias." *Economics Letters* 108, no. 2 (2010): 212–14.

Pope, Devin G., and Maurice E. Schweitzer. "Is Tiger Woods Loss Averse? Persistent Bias in the Face of Experience, Competition, and High Stakes." *American Economic Review* 101, no. 1 (2011): 129–57.

Prendergast, Canice. "The Provision of Incentives in Firms." *Journal of Economic Literature* 37, no. 1 (1999): 7–63.

——. "The Tenuous Trade-off Between Risk and Incentives." *Journal of Political Economy* 110, no. 5 (2002): 1071–102.

Price, Joseph, and Justin Wolfers. "Racial Discrimination among NBA Referees." *Quarterly Journal of Economics* 125, no. 4 (2010): 1859–87.

Priks, Mikael. "Hooliganomics." Master's thesis, Institute for International Economic Studies, Stockholm University, Sweden, May 2008.

Prouhet, Eugene. "Memoire sur Quelques Relations entre les Puissances des Nombres." *Comptes Rendus des Scéances de l'Académie des Sciences. Paris* 33 (1851): 225–26.

Rabin, Matthew. "Inference by Believers in the Law of Small Numbers." *Quarterly Journal of Economics* 117, no. 3 (2002): 775–816.

——. "Psychology and Economics." *Journal of Economic Literature* 36, no.1 (1998): 11–46.

Rapoport, Amnon, and Richard B. Boebel. "Mixed Strategies in Strictly Competitive Games: A Further Test of the Minimax Hypothesis." *Games and Economic Behavior* 4, no. 2 (1992): 261–83.

Rapoport, Amnon, and David V. Budescu. "Generation of Random Sequences in Two-Person Strictly Competitive Games." *Journal of Experimental Psychology: General* 121, no. 3 (1992): 352–63.

Rauh, Michael T., and Giulio Seccia. "Anxiety and Performance: An Endogenous Learning-by-Doing Model." *International Economic Review* 47, no. 2 (2006): 583–609.

Relaño, Alfredo. *366 Historias del Fútbol Mundial que Deberías Leer*. Madrid: Ediciones Martínez Roca, 2010.

Richman, Robert. "Recursive Binary Sequences of Differences." *Complex Systems* 13, no. 4 (2001): 381–92.

Rickman, Neil, and Robert Witt. "Favouritism and Financial Incentives: A Natural Experiment." *Economica* 75, no. 298 (2008): 296–309.

Rob, Rafael, and Peter Zemsky. "Social Capital, Corporate Culture and Incentive Intensity." *RAND Journal of Economics* 33, no. 2 (2002): 243–57.

Rosen, Sherwin H. "Prizes and Incentives in Elimination Tournaments." *American Economic Review* 76, no. 4 (1986): 701–15.

Rosen, Sherwin H., and Allen Sanderson. "Labour Markets in Professional Sports." *The Economic Journal* 111, no. 469 (2001): 47–68.

Rubinstein, Ariel. "Afterword." In 60th anniversary ed. of *Theory of Games and Economic Behavior* by Oskar Morgenstern and John von Neumann, 633–36. Princeton, NJ: Princeton University Press, 2004.

——. *Economic Fables*. Cambridge, UK: Open Book Publishers, 2012.

Sala-i-Martin, Xavier. *Pues Yo Lo Veo Así*. Barcelona: Random House Mondadori, 2010.

Schelling, Thomas C. "Dynamic Models of Segregation." *Journal of Mathematical Sociology* 1, no. 2 (1971): 143–86.

Seneca, Lucius Annaeus. *Letters from a Stoic (Epistolae Morales ad Lucilium)*. London: Penguin Books, 1969.

Seo, H., D. J. Barraclough, and Daeyol Lee. "Lateral Intraparietal Cortex and Reinforcement Learning During a Mixed-Strategy Game." *Journal of Neuroscience* 29 (2009): 7278–89.

Smith, Adam. *The Theory of Moral Sentiments,* edited by D. D. Raphael and A. L. Macfie, London: Oxford University Press, 1976.

Szymanski, Stefan. "A Market Test for Discrimination in the English Professional Soccer Leagues." *Journal of Political Economy* 108, no. 3 (2000): 590–603.

——. *Playbooks and Checkbooks*. Princeton, NJ: Princeton University Press, 2009.

Thue, Axel. "Über die gegenseitige Lage gleicher Teile gewisser Zeichenreihen." *Christiana Videnskabs-Selskabs Skrifter. I. Math.-Naturv. Klasse* 1 (1912). Reprinted in *Selected Mathematical Papers of Axel Thue,* edited by T. Nagell. Oslo: Universitetsforlaget, 1977, 139–58.

Tversky, Amos, and Daniel Kahneman. "The Belief in the Law of Small Numbers." *Psychological Bulletin* 76, no. 2 (1971): 105–10.

*USA Today.* "FIFA Officials' Goal: Encourage Attacking, High-Scoring Matches." March 17, 1994.

Vaughan Williams, Leighton. *Information Efficiency in Financial and Betting Markets.* Cambridge, UK: Cambridge University Press, 2005.

——. *Prediction Markets: Theory and Application.* London: Routledge International Studies in Money and Banking, 2011.

Vickery, Timothy J., Marvin M. Chun, and Daeyeol Lee. "Ubiquity and Specificity of Reinforcement Signals throughout the Human Brain." *Neuron* 72 (October 6, 2011): 166–77.

Vickery, Timothy J., and Yuhong V. Jiang. "Inferior Parietal Lobule Supports Decision Making under Uncertainty in Humans." *Cerebral Cortex* 19 (April 2009): 916–25.

von Neumann, John. "Communication on the Borel Notes." *Econometrica* 21, no. 1 (1953): 124–27.

——. "The Mathematician." In *The Works of the Mind.* Chicago: University of Chicago Press, 1947: 180–96.

——. "A Model of General Equilibrium." *Review of Economic Studies* 13, no. 1 (1945–1946): 1-9.

——. "Über ein okonomisches Gleichungssystem und eine Verallgemeinerung des Brouwerschen Fixpunktsatzes." *Ergebnisse Eines Mathematischen Kolloquiums* 8 (1937): 73–83.

——. Zur Theorie der Gesellschaftsspiele." *Mathematische Annalen* 100 (1928): 295–300.

von Neumann, John, and Oskar Morgenstern. *Theory of Games and Economic Behavior.* Princeton, NJ: Princeton University Press, 1944.

Walker, Mark, and John Wooders. "Minimax Play at Wimbledon." *American Economic Review* 91, no. 5 (2001): 1521–38.

WMD Commission Report. "Intelligence Capabilities of the United States Regarding Weapons of Mass Destruction." Washington, DC: Commission on the Intelligence Capabilities of the United States Regarding Weapons of Mass Destruction, March 2005.

Wolfers, Justin, and Eric Zitzewitz. "Interpreting Prediction Market Prices as Probabilities." London: Centre for Economic Policy Research, CEPR Discussion Paper 5676, 2006.

——. "Prediction Markets." *Journal of Economic Perspectives* 18, no. 2 (2004): 107–26.

Wright, John F. "British Government Borrowing in Wartime, 1750–1815." *The Economic History Review* 52, no. 2 (1999): 355–61.

Yannis, Alex. "Avoid Scoreless Ties." *New York Times,* January 4, 1994.

# INDEX